World Flags

Afghanistan · Albania · Algeria · Andorra · Angola · Antigua and Barbuda · Argentina

Armenia · Australia · Austria · Azerbaijan · Bahamas · Bahrain · Bangladesh

Barbados · Belarus · Belgium · Belize · Benin · Bhutan · Bolivia

Bosnia-Herzegovina · Botswana · Brazil · Brunei · Bulgaria · Burkina · Burundi

Cambodia · Cameroon · Canada · Cape Verde · Central African Republic · Chad · Chile

China · Colombia · Comoros · Congo · Congo, Dem. Rep. · Costa Rica · Côte d'Ivoire

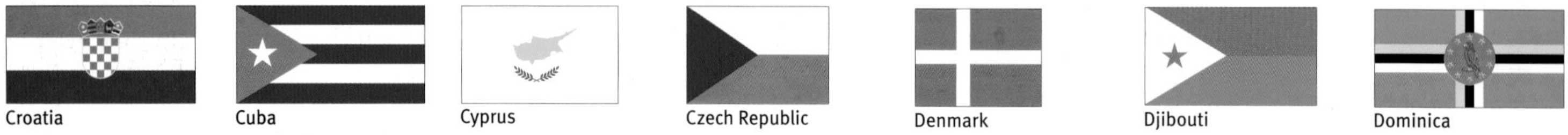
Croatia · Cuba · Cyprus · Czech Republic · Denmark · Djibouti · Dominica

Dominican Republic · East Timor · Ecuador · Egypt · El Salvador · Equatorial Guinea · Eritrea

Estonia · Ethiopia · Fiji · Finland · France · French Guiana · Gabon

Gambia · Georgia · Germany · Ghana · Greece · Greenland · Grenada

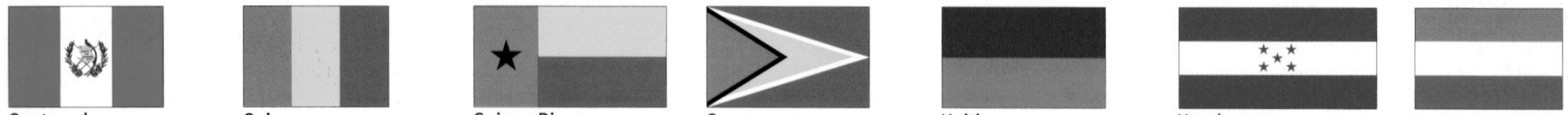
Guatemala · Guinea · Guinea-Bissau · Guyana · Haiti · Honduras · Hungary

Iceland · India · Indonesia · Iran · Iraq · Ireland · Israel

Italy · Jamaica · Japan · Jordan · Kazakhstan · Kenya · Kiribati

Kuwait · Kyrgyzstan · Laos · Latvia · Lebanon · Lesotho · Liberia

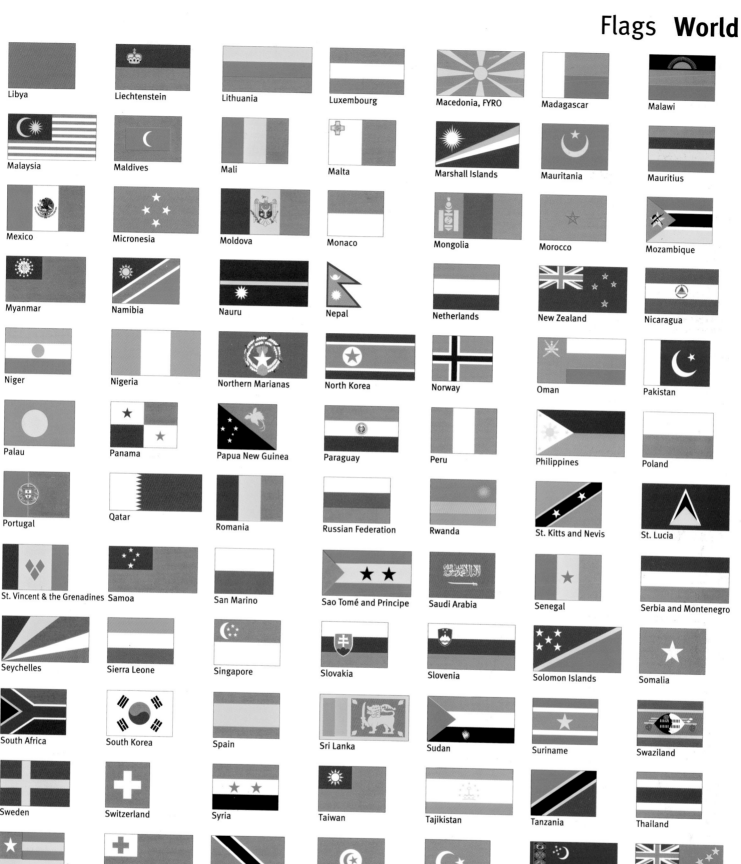

Libya

Liechtenstein

Lithuania

Luxembourg

Macedonia, FYRO

Madagascar

Malawi

Malaysia

Maldives

Mali

Malta

Marshall Islands

Mauritania

Mauritius

Mexico

Micronesia

Moldova

Monaco

Mongolia

Morocco

Mozambique

Myanmar

Namibia

Nauru

Nepal

Netherlands

New Zealand

Nicaragua

Niger

Nigeria

Northern Marianas

North Korea

Norway

Oman

Pakistan

Palau

Panama

Papua New Guinea

Paraguay

Peru

Philippines

Poland

Portugal

Qatar

Romania

Russian Federation

Rwanda

St. Kitts and Nevis

St. Lucia

St. Vincent & the Grenadines

Samoa

San Marino

Sao Tomé and Principe

Saudi Arabia

Senegal

Serbia and Montenegro

Seychelles

Sierra Leone

Singapore

Slovakia

Slovenia

Solomon Islands

Somalia

South Africa

South Korea

Spain

Sri Lanka

Sudan

Suriname

Swaziland

Sweden

Switzerland

Syria

Taiwan

Tajikistan

Tanzania

Thailand

Togo

Tonga

Trinidad and Tobago

Tunisia

Turkey

Turkmenistan

Tuvalu

Uganda

Ukraine

United Arab Emirates

United Kingdom

United States of America

Uruguay

Uzbekistan

Vanuatu

Venezuela

Vietnam

Yemen

Zambia

Zimbabwe

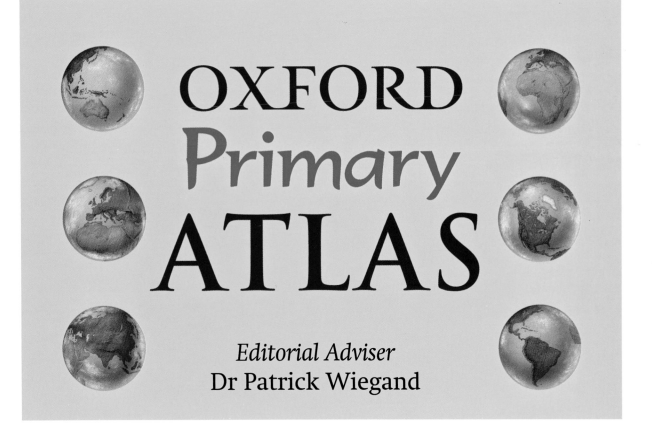

OXFORD Primary ATLAS

Editorial Adviser
Dr Patrick Wiegand

OXFORD
UNIVERSITY PRESS

Great Clarendon Street, Oxford OX2 6DP

Oxford University Press is a department of the University of Oxford.
It furthers the University's objective of excellence in research, scholarship,
and education by publishing worldwide in

Oxford   New York

Auckland   Cape Town   Dar es Salaam   Hong Kong   Karachi
Kuala Lumpur   Madrid   Melbourne   Mexico City   Nairobi
New Delhi   Shanghai   Taipei   Toronto

With offices in

Argentina   Austria   Brazil   Chile   Czech Republic   France   Greece
Guatemala   Hungary   Italy   Japan   Poland   Portugal   Singapore
South Korea   Switzerland   Thailand   Turkey   Ukraine   Vietnam

Oxford is a registered trade mark of Oxford University Press
in the UK and in certain other countries

ISBN 0 19 832160 0 (hardback)

ISBN 0 19 832159 7 (paperback)

3 5 7 9 10 8 6 4 2

Printed in Singapore

Acknowledgements

Illustrations by:
Julian Baker p 20; Adrian Barclay pp 31, 35, 41, 51, 54, 55; Mark Duffin pp 7 (compass), 23tl, tr & br, 24 (bricks), 26, 59; Nick Hawken pp 24 (settlements), 32, 33, 36, 37, 38, 39, 42, 43, 45, 46, 47, 48, 49, 52, 53, 55; Tracey Learoyd and Adrian Smith p 20 *and thereafter* (landscape pictograms); ODI p 24 *and thereafter* (population figures); Harry Venning p 60

The publishers would also like to thank the following for permission to reproduce the following photographs:
Alamy pp 20t (Robert Harding Picture Library), 20ct (The Photolibrary Wales), 20c (Geogphotos), 20cb (Worldwide Picture Library), 22tr (Leslie Garland Picture Library), 24t (David Crausby), 24c (Elmtree Images), 26t (David Martyn Hughes), 26b (Gina Calvi), 29ct (Jon Arnold Images), 29cb (Ian Thraves), 39t (Robert Harding Picture Library), 48t (ImageState), 60cr (Steve Bloom Images), 60l (TH Foto), 60r (Guy Somerset), 64tc (Robert Harding Picture Library), 64bl (ashfordplatt); Corbis pp 20b (Chinch Gryniewicz), 22tl (David Paterson), 23tl (Neil Beer), 26c (Jason Hawkes), 28t (Martin Jones), 33b (ML Sinibaldi), 36t (Lindsay Hebberd), 36b (Richard Bickel), 42br (Charles Lenars), 52t (Jeremy Horner), 64tr (Ron Watts), 65tr (Richard A. Cooke), 65bc (Galen Rowell), 65bl (Wolfgang Kaehler); Frank Lane Picture Agency pp 29b (Chris Demetriou), 42br (Derek Hall), 43ct (Peter Davey), 43br (David Hosking), 64tl (Minden Pictures); Getty Images/Photographer's Choice p 46l (James Randklev); Getty Images/Stone pp 24b (Patrick Ingrand), 29tr (Tony Page), 33t, 42t (Will & Deni McIntyre), 43l (Daryl Balfour), 46r, 52b (Pascal Rondeau), 58br; Getty Taxi pp 28b (Richard Cooke), 39b, 47r (B & M Productions), 48b (Tom Bean); Getty Images/The Image Bank pp 47l, 58bl (Image Makers), 64br, 65tl, 65br (Frans Lemmens); Heritage Image Partnership © The British Museum p 28ct (Institution Reference: M&ME, 1939,10-10,93); Powerstock p 28cb (Superstock); Science Photo Library pp 8 (NRSC Ltd), 59l (Earth Satellite Corporation), 59r (Planetary Visions Ltd), 60cl (David Vaughan); © UK Perspectives p 27.

The page design is by Adrian Smith.

The publishers are grateful to the following colleagues in geography education for their helpful comments and advice during the development stages of this atlas:

Jeremy Bullock, Susan Butler, Claire Condie, John Dewis, Tracey Ellis, John Halocha, Joan Huckle, Richard Jefferies, Pat Kelway, Amanda Lightfoot, Trevor Mason, Vanessa Richards, Vicky Stevenson, Emma Wells, Niki Whitburn, Brenda Whittle.

The publishers would also like to thank Phoenix Mapping and Suzanne Williams for their help during the production of this atlas.

2 Contents

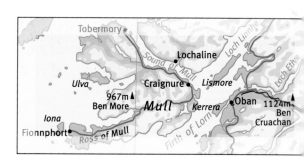

The United Kingdom

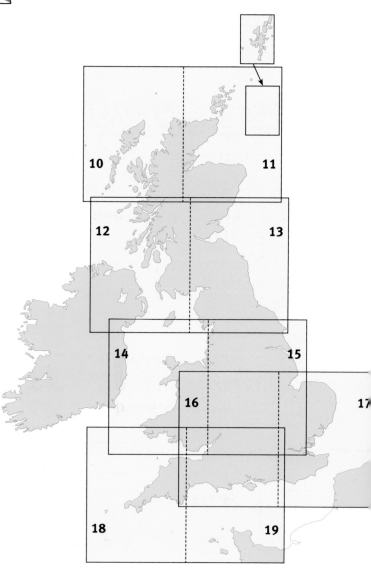

© Oxford University Press

Contents 3

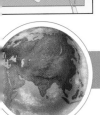

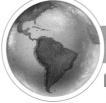
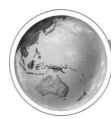

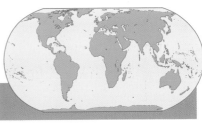

4 Atlas literacy

Map language

There are special names for the parts of maps

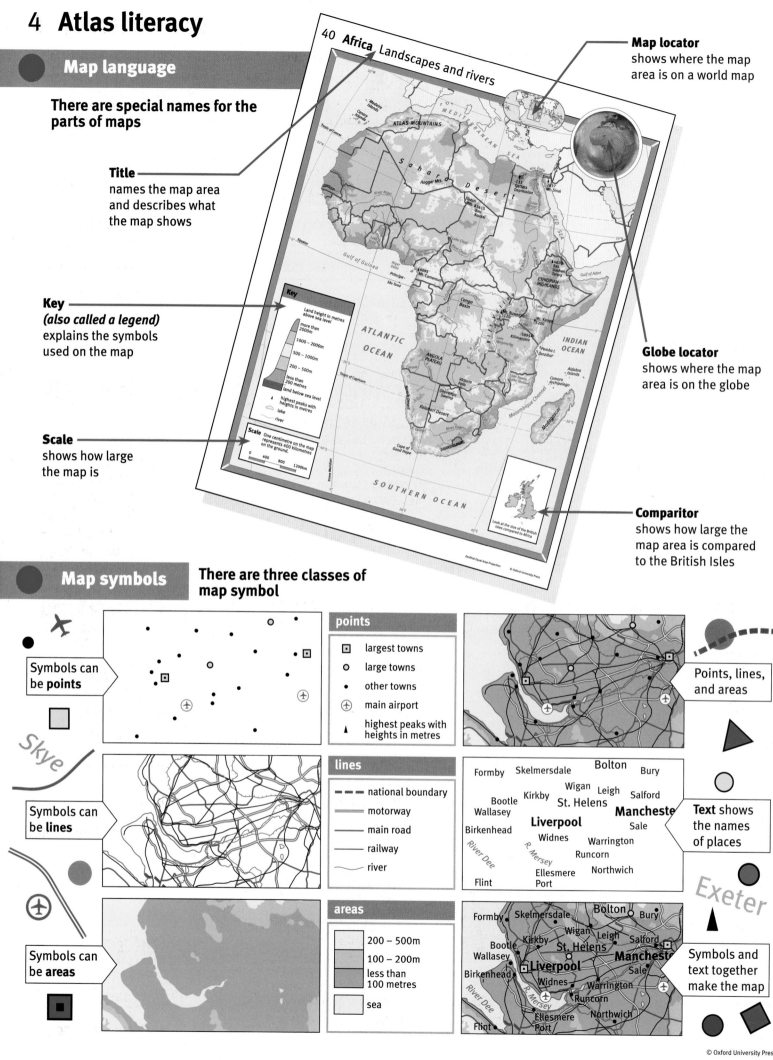

Title
names the map area and describes what the map shows

Key
(also called a legend)
explains the symbols used on the map

Scale
shows how large the map is

Map locator
shows where the map area is on a world map

Globe locator
shows where the map area is on the globe

Comparitor
shows how large the map area is compared to the British Isles

40 **Africa** Landscapes and rivers

Map symbols

There are three classes of map symbol

Symbols can be **points**

Symbols can be **lines**

Symbols can be **areas**

Skye

Exeter

points

⊡	largest towns
○	large towns
•	other towns
⊕	main airport
▲	highest peaks with heights in metres

lines

- - -	national boundary
═══	motorway
───	main road
───	railway
───	river

areas

	200 – 500m
	100 – 200m
	less than 100 metres
	sea

Points, lines, and areas

Text shows the names of places

Symbols and text together make the map

Formby Skelmersdale Bolton Bury
Wigan Leigh Salford
Bootle Kirkby St. Helens
Wallasey **Manchester**
Birkenhead **Liverpool** Sale
Widnes Warrington
River Dee R. Mersey Runcorn
Ellesmere Northwich
Flint Port

© Oxford University Press

Type on maps

The way text is printed on maps gives an important clue to what the words mean

Great Britain *Ireland*	islands
UNITED KINGDOM REPUBLIC OF IRELAND	countries
ENGLAND SCOTLAND WALES NORTHERN IRELAND	parts of the United Kingdom
PENNINES GRAMPIAN MOUNTAINS	physical features
Ben Nevis Snowdon	mountain peaks
NORTH SEA *English Channel*	sea areas
Manchester York Dover	settlements

Map abbreviations

An abbreviation is a shortened version of a word or a group of words

Some country names are abbreviated using the first letters of each word

R.	River
Mt.	Mount
Is.	Island
Pen.	Peninsula

UK   United Kingdom
USA United States of America
UAE United Arab Emirates

Country names and adjectives

There are patterns in the way some country names make adjectives

Australia	Australian
India	Indian
Nigeria	Nigerian
Zambia	Zambian

Ireland	Irish
Poland	Polish
Sweden	Swedish
Turkey	Turkish

China	Chinese
Japan	Japanese
Malta	Maltese
Taiwan	Taiwanese

Brazil	Brazilian
Canada	Canadian
Egypt	Egyptian
Italy	Italian

Bangladesh	Bangladeshi
Iraq	Iraqi
Israel	Israeli
Pakistan	Pakistani

Other country names make adjectives with no pattern

Cyprus	Cypriot
France	French
Germany	German

Greece	Greek
Iceland	Icelandic
Netherlands	Dutch

Peru	Peruvian
Slovakia	Slovak
Thailand	Thai

Map punctuation

A country name in brackets shows that a place is part of that country

Corsica is part of France

The Bulgarian *flag*

The Danish *flag*

The Argentinian *flag*

The Senegalese *flag*

6 Atlas numeracy

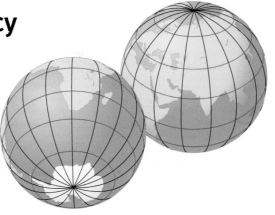

The Earth is a sphere*.

Two sets of imaginary lines help us describe where places are on the Earth.

All the lines are numbered and some have special names.

* It's actually slightly flattened at the north and south poles.

Longitude

Lines of longitude measure distance east or west of the Prime Meridian.

The **Prime Meridian** (also called the Greenwich Meridian) is at longitude 0°.

The **International Date Line** (on the other side of the Earth) is based on longitude 180°.

Latitude

Lines of latitude measure distance north or south of the equator.

The equator is at latitude 0°.

The poles are at latitude 90°N and 90°S.

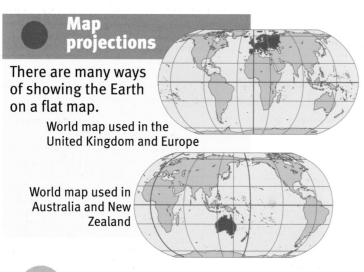

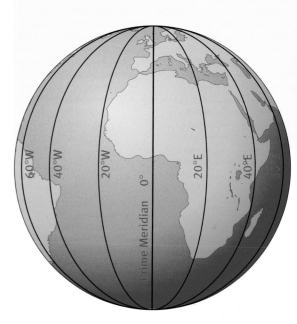

Map projections

There are many ways of showing the Earth on a flat map.

World map used in the United Kingdom and Europe

World map used in Australia and New Zealand

Grid codes

In this atlas, the lines of latitude and longitude are used to make a grid.

The columns of the grid have letters.

The rows of the grid have numbers.

Numbers and letters together make a grid code that can be used to describe where places are on the Earth.

Abuja is in B4   Durban is in C2

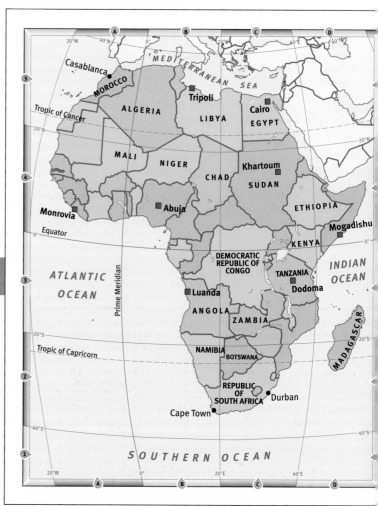

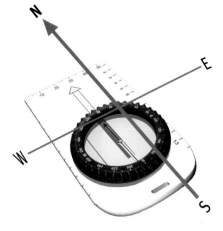

A compass is used for finding direction.

The needle of a compass always points north.

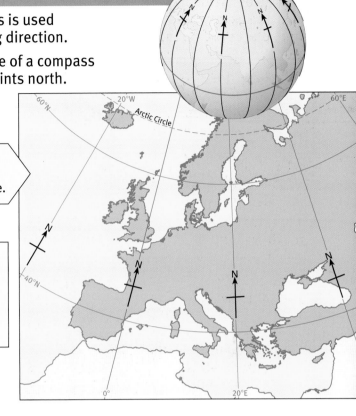

North on atlas maps follows the lines of longitude.

London is north of Brighton.

Brighton is south of London.

Reading is west of London.

Portsmouth is south west of London.

Scale

Maps are much, much smaller than the countries they show.

A few centimetres on the map stand for very many kilometres on the ground.

Each division on the scale line is one centimetre. The scale line shows how many kilometres are represented by one centimetre.

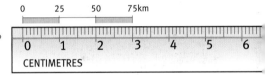

CENTIMETRES

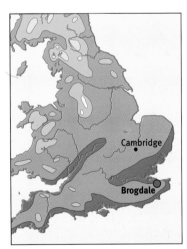

Scale One centimetre on the map represents **25** kilometres on the ground.

The distance between Bangor and Betws-y-Coed is about 25km

Scale One centimetre on the map represents **50** kilometres on the ground.

The distance between Perth and Edinburgh is about 50km

Scale One centimetre on the map represents **100** kilometres on the ground.

The distance between Cambridge and Brogdale is about 100km

Larger scale smaller area more detail

Smaller scale larger area less detail

On world maps the scale is only true along the equator.

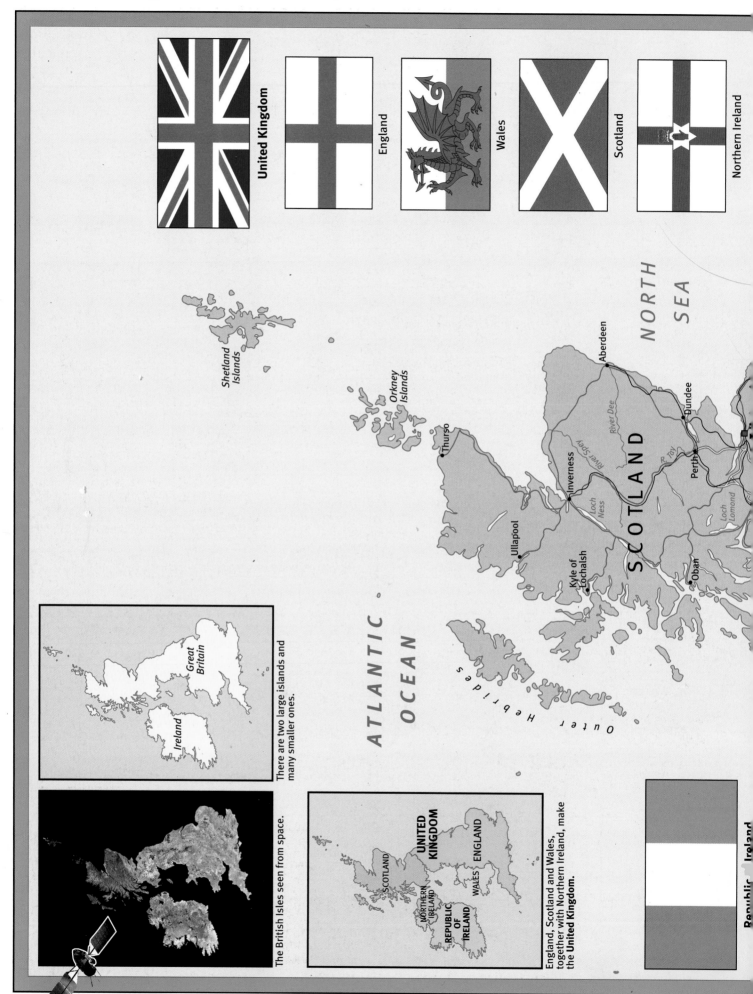

United Kingdom

England

Wales

Scotland

Northern Ireland

Shetland Islands

Orkney Islands

ATLANTIC OCEAN

NORTH SEA

Aberdeen

Dundee

River Dee

River Spey

R. Tay

Perth

SCOTLAND

Inverness

Loch Ness

Loch Lomond

Thurso

Ullapool

Kyle of Lochalsh

Oban

Outer Hebrides

Great Britain

Ireland

There are two large islands and many smaller ones.

The British Isles seen from space.

UNITED KINGDOM

ENGLAND

WALES

SCOTLAND

NORTHERN IRELAND

REPUBLIC OF IRELAND

England, Scotland and Wales, together with Northern Ireland, make the **United Kingdom.**

Republic of Ireland

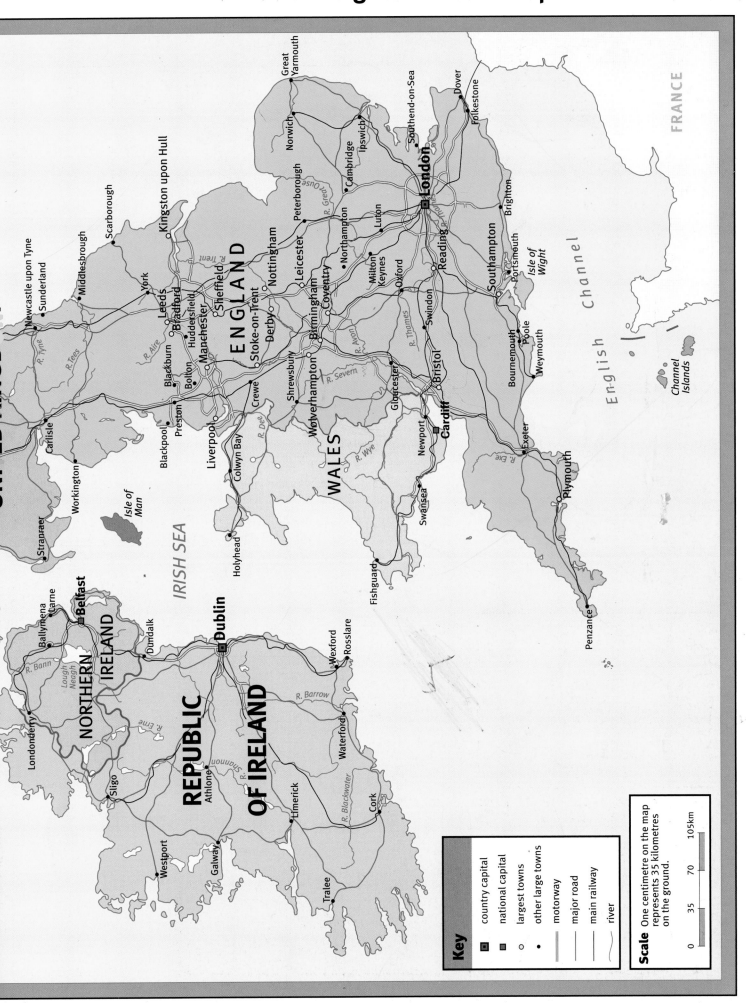

Key

- ◼ country capital
- ◼ national capital
- ○ largest towns
- • other large towns
- motorway
- major road
- main railway
- river

Scale One centimetre on the map represents 35 kilometres on the ground.

0   35   70   105km

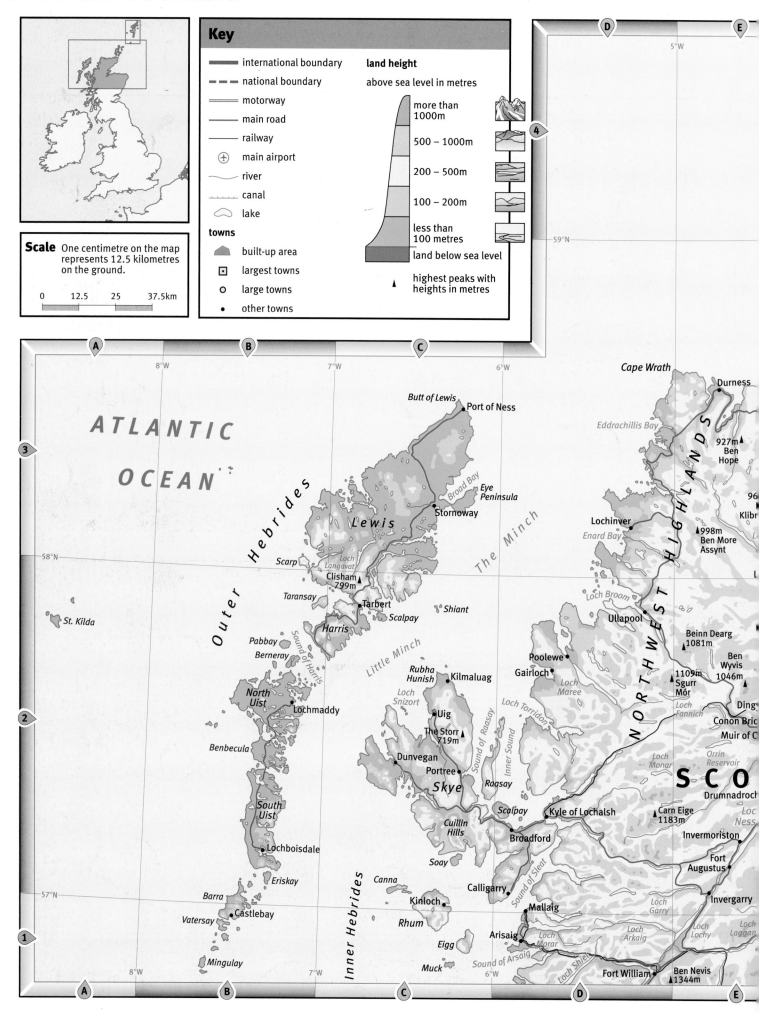

Key

▬▬▬	international boundary
▬ ▬ ▬	national boundary
══	motorway
──	main road
──	railway
⊕	main airport
⌒	river
⊤⊤⊤	canal
◠	lake

towns

⬢	built-up area
⊡	largest towns
○	large towns
•	other towns

land height

above sea level in metres

- more than 1000m
- 500 – 1000m
- 200 – 500m
- 100 – 200m
- less than 100 metres
- land below sea level

▲ highest peaks with heights in metres

Scale One centimetre on the map represents 12.5 kilometres on the ground.

0   12.5   25   37.5km

ATLANTIC OCEAN

Cape Wrath
Durness
Eddrachillis Bay
927m Ben Hope
Butt of Lewis
Port of Ness
Broad Bay
Eye Peninsula
Stornoway
The Minch
Lochinver
Enard Bay
998m Ben More Assynt
Outer Hebrides
Lewis
Loch Langavat
Scarp
Clisham 799m
Taransay
Tarbert
Scalpay
Shiant
Harris
Loch Broom
Loch Maree
Ullapool
Beinn Dearg 1081m
Ben Wyvis 1046m
1109m Sgurr Mór
Loch Fannich
Ding
Conon Bri
Muir of C
Pabbay
Berneray
Sound of Harris
Little Minch
Rubha Hunish
Kilmaluag
Poolewe
Gairloch
Loch Torridon
St. Kilda
North Uist
Lochmaddy
Loch Snizort
Uig
The Storr 719m
Dunvegan
Portree
Skye
Sound of Raasay
Inner Sound
Raasay
Loch Monar
Orrin Reservoir
SCO
Drumnadroch
Benbecula
South Uist
Lochboisdale
Cuillin Hills
Scalpay
Kyle of Lochalsh
Carn Eige 1183m
Loch Ness
Invermoriston
Broadford
Soay
Inner Hebrides
Canna
Eriskay
Barra
Castlebay
Vatersay
Kinloch
Rhum
Eigg
Muck
Calligarry
Sound of Sleat
Mallaig
Arisaig
Loch Morar
Sound of Arsaig
Loch Shiel
Fort William
Ben Nevis 1344m
Fort Augustus
Invergarry
Loch Garry
Loch Arkaig
Loch Lochy
Loch Laggan
Mingulay
NORTH WEST HIGHLANDS

Main map (left):

F — G — 2°W

4°W 3°W 2°W

Fair Isle

4

Mull Head
Papa
Westray

North
Ronaldsay

Westray

Sanday

Westray Firth

Rousay

Eday

Brough Head

Stronsay

Stronsay Firth

Mainland

Shapinsay

Orkney
Islands

59°N

Kirkwall

Stromness

Scapa

479m ▲
Ward Hill

Scapa
Flow

Hoy

South Ronaldsay

Pentland Firth

3

Dunnet
Head

Stroma

Duncansby
Head

John
o'Groats

Thurso

...tyhill

Halkirk

River Wick

Wick

River Thurso

Lybster

Loch
Nan
Clár

Kinbrace

Helmsdale

Brora

Golspie

58°N

Dornoch Firth Tarbat Ness

Tain

Invergordon

...marty

Moray Firth

Lossiemouth

Portknockie

Cullen

Rosehearty

Fraserburgh

Elgin

Portsoy

Macduff

Nairn

Fochabers

Forres

Keith

River Deveron

Turriff

Mintlaw

Peterhead

Rothes

Huntly

Dufftown

NORTH

SEA

2

Oldmeldrum

Ellen

...verness

AND

Grantown-on-Spey

Inverurie

River Don

...adhliath
...untains

Aviemore

Dyce

Aberdeen

...gussie

Cairngorms

1244m ▲
Cairn Gorm

Aboyne

River Dee

57°N

Newtonmore

Ballater

Lochnagar
1155m

Banchory

Braemar

...alwhinnie

AMPIAN MOUNTAINS

R. North Esk

Stonehaven

1

Inverbervie

4°W 3°W 2°W

F — G — H — J

Inset map (right): Shetland Islands

G — H — J

2°W 1°W

Herma Ness

Haroldswick

Unst

Point of
Fethaland

Yell Sound

Yell

Fetlar

Esha Ness

Out
Skerries

5 St. Magnus Bay

Muckle
Roe

Whalsay

Papa
Stour

Mainland

417m ▲
Foula

Walls

Bressay

Scalloway

Lerwick

Shetland
Islands

60°N 60°N

Sumburgh
Head

4

2°W 1°W

Fair Isle

G — H — J

Scale One centimetre on the map represents 12.5 kilometres on the ground.

0   12.5   25   37.5km

© Oxford University Press

Transverse Mercator Projection

Key

▬▬▬	international boundary
▬ ▬ ▬	national boundary
▭▭	motorway
──	main road
──	railway
✈	main airport
～	river
┈┈	canal
◠	lake

towns

⬢	built-up area
⊡	largest towns
○	large towns
•	other towns

land height

above sea level in metres

	more than 1000m	
	500 – 1000m	
	200 – 500m	
	100 – 200m	
	less than 100 metres	
	land below sea level	
▲	highest peaks with heights in metres	

IRISH SEA

Dundalk Bay

- Warrenpoint
- Kilkeel
- Balbriggan
- Swords
- Dublin
- Dún Laoghaire
- Bray
- Wicklow
- Arklow

Cahore Point

St. George's Channel

Port Erin
Calf of Man
Castletown
Isle of Man

Carmel Head
Amlwch
Holyhead
Anglesey
Holy Island
Llangefni
Menai Bridge
Bangor
Bethesda
Caernarfon
Caernarfon Bay

Llandudno
Conwy
Rhyl
Colwyn Bay
Denbigh
Llanrwst
Betws-y-Coed
▲ Snowdon 1085m

Barrow-in-Furness
Morecambe
Lancaster
Heysham
Carnforth

Fleetwood
Thornton
Blackpool
Preston
Lytham St. Anne's
Leyland
Southport
Formby
Skelmersdale
Bootle
Kirkby
Wallasey
Birkenhead
Liverpool
Widnes
Runco
Ellesmere Port
Winsf
Chester
Wrexham
Nantwi

Prestatyn
Holywell
Connah's Quay
Flint
Mold
Ruthin
River Dee
R. Mersey

Lleyn Peninsula
Porthmadog
Pwllheli
Harlech
Bardsey Island

Blaenau Ffestiniog
Bala
Bala Lake
Corwen
Llangollen

Dolgellau
▲ 892m Cadair Idris
Barmouth
Tywyn
Machynlleth
R. Dyfi

CAMBRIAN MOUNTAINS

Lake Brenig
Lake Vyrnwy
R. Dee
Ruabon
Chirk
Oswestry
Shrewsbury
Welshpool
Montgomery
Wenlock Edge
The W

Cardigan Bay

Plynlimon 752m
Newtown
Llanidloes
Aberystwyth

W A L E S

Aberaeron
New Quay
Lampeter

Cemaes Head
Cardigan
River Teifi
Newport
Newcastle Emlyn

Strumble Head
Fishguard
Preseli Mountains

St. David's Head
St. Brides Bay

Carmarthen
St. Clears
Milford Haven
Pembroke Dock
Pembroke
Saundersfoot
Tenby
Carmarthen Bay

Rhayader
Claerwen Reservoir
Lake Brianne Reservoir
Llandrindod Wells
Builth Wells

Knighton
Kington
Ludlo
River
Leomi

Llandovery
R. Tywi
Llandeilo
Ammanford
Cross Hands
Kidwelly
Burry Port
Llanelli
Pontardulais
Gorseinon
Gower
Worms Head
Swansea
Neath
Maesteg
Port Talbot
Porthcawl
Bridgend
Barry
Penarth

Mynydd Eppynt
River Usk
Brecon
Brecon Beacons
▲ 886m
Abergavenny
Tredegar
Ebbw Vale
Abertillery
Merthyr Tydfil
Aberdare
Mountain Ash
Rhondda
Gelligaer
Pontypool
Cwmbran
Pontypridd
Caerphilly
Cardiff
Clevedon

Black Mountains
Hay-on-Wye
Heref
Ross-on-Wye
Monmouth
Cin
R. Wye
Chepstow
Newport
Mangotsf
Bristol
Kings

R. Neath

Bristol Channel
Weston-super-Mare
Keyr

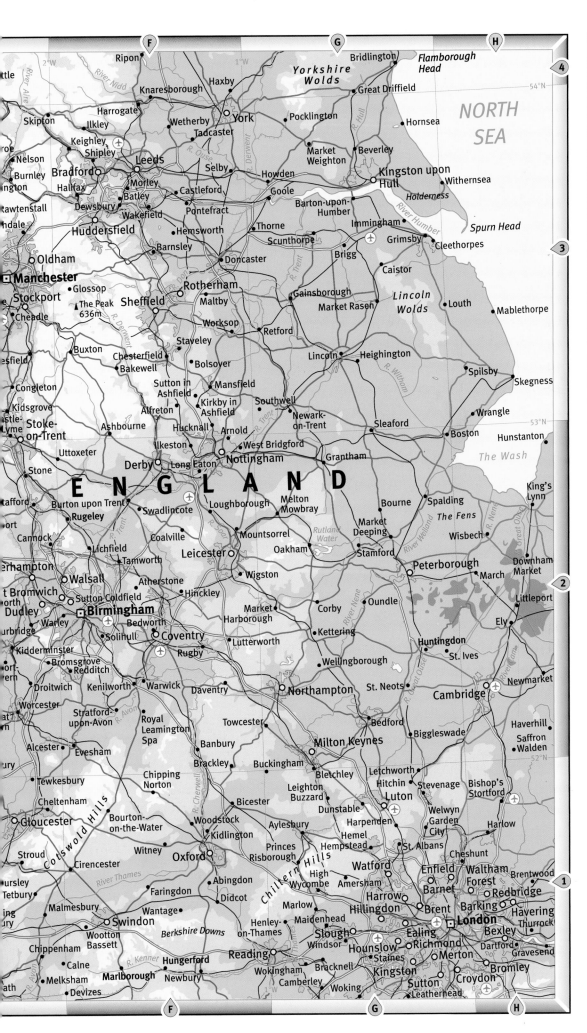

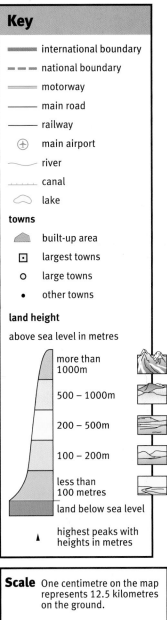

Key

- ▬▬▬ international boundary
- ▬ ▬ ▬ national boundary
- ═══ motorway
- ▬▬▬ main road
- ▬▬▬ railway
- ⊕ main airport
- ～～ river
- ⊥⊥⊥ canal
- ⬠ lake

towns

- 🔺 built-up area
- ⊡ largest towns
- ○ large towns
- • other towns

land height

above sea level in metres

- more than 1000m
- 500 – 1000m
- 200 – 500m
- 100 – 200m
- less than 100 metres
- land below sea level

- ▲ highest peaks with heights in metres

Scale One centimetre on the map represents 12.5 kilometres on the ground.

0   12.5   25   37.5km

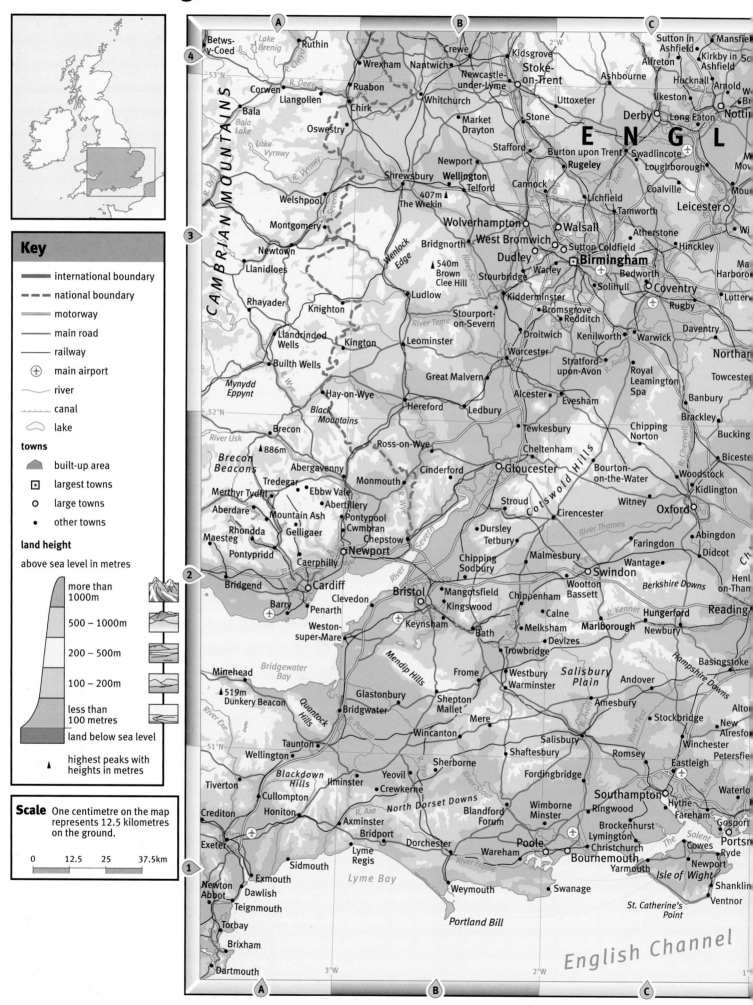

Key

▰▰▰	international boundary
▰ ▰ ▰	national boundary
══	motorway
──	main road
──	railway
✈	main airport
∿	river
┼┼┼	canal
⌒	lake

towns

⬟	built-up area
⊡	largest towns
○	large towns
•	other towns

land height

above sea level in metres

	more than 1000m
	500 – 1000m
	200 – 500m
	100 – 200m
	less than 100 metres
	land below sea level
▲	highest peaks with heights in metres

Scale One centimetre on the map represents 12.5 kilometres on the ground.

0   12.5   25   37.5km

The Wash

The Fens

NORTH SEA

North Downs

South Downs

Strait of Dover

FRANCE

BELGIUM

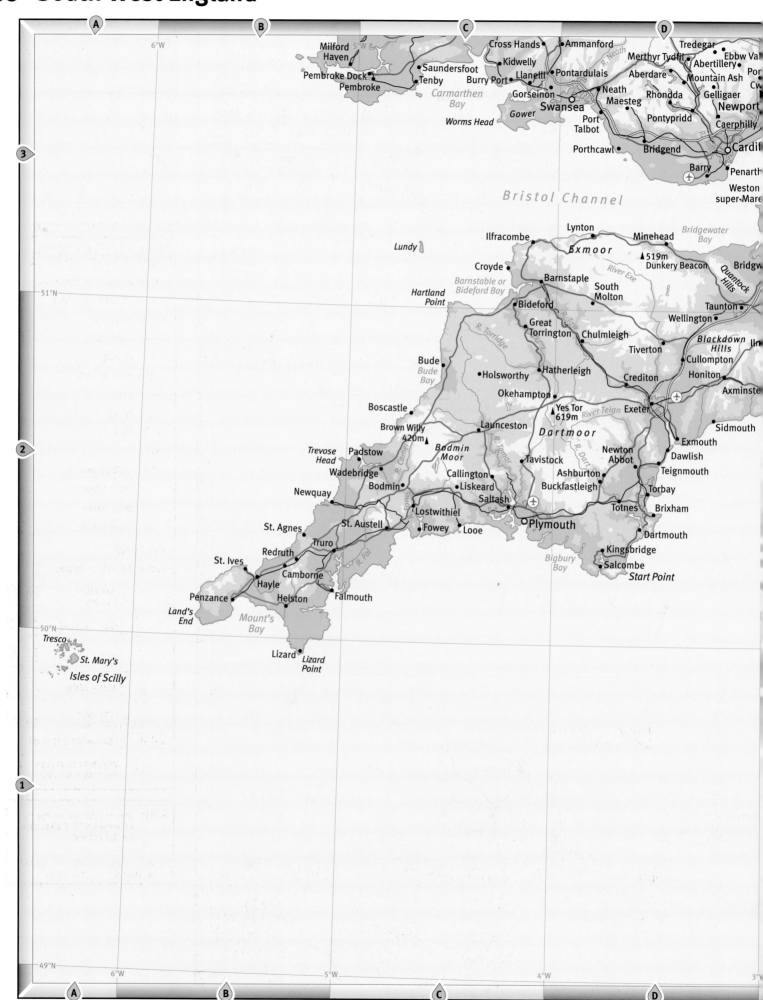

Key

	international boundary
	national boundary
	motorway
	main road
	railway
⊕	main airport
	river
	canal
	lake

towns

	built-up area
⊡	largest towns
○	large towns
•	other towns

land height

above sea level in metres

more than 1000m	
500 – 1000m	
200 – 500m	
100 – 200m	
less than 100 metres	
land below sea level	
▲	highest peaks with heights in metres

Scale One centimetre on the map represents 12.5 kilometres on the ground.

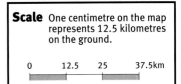

0   12.5   25   37.5km

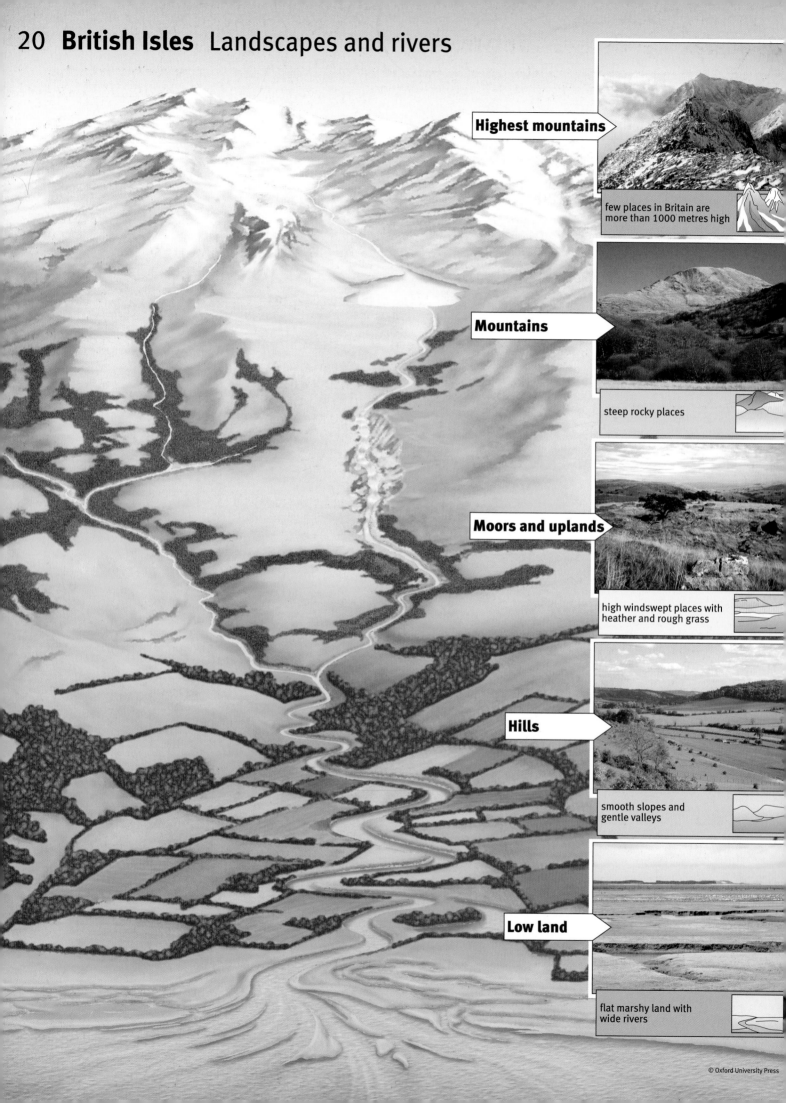

Highest mountains

few places in Britain are more than 1000 metres high

Mountains

steep rocky places

Moors and uplands

high windswept places with heather and rough grass

Hills

smooth slopes and gentle valleys

Low land

flat marshy land with wide rivers

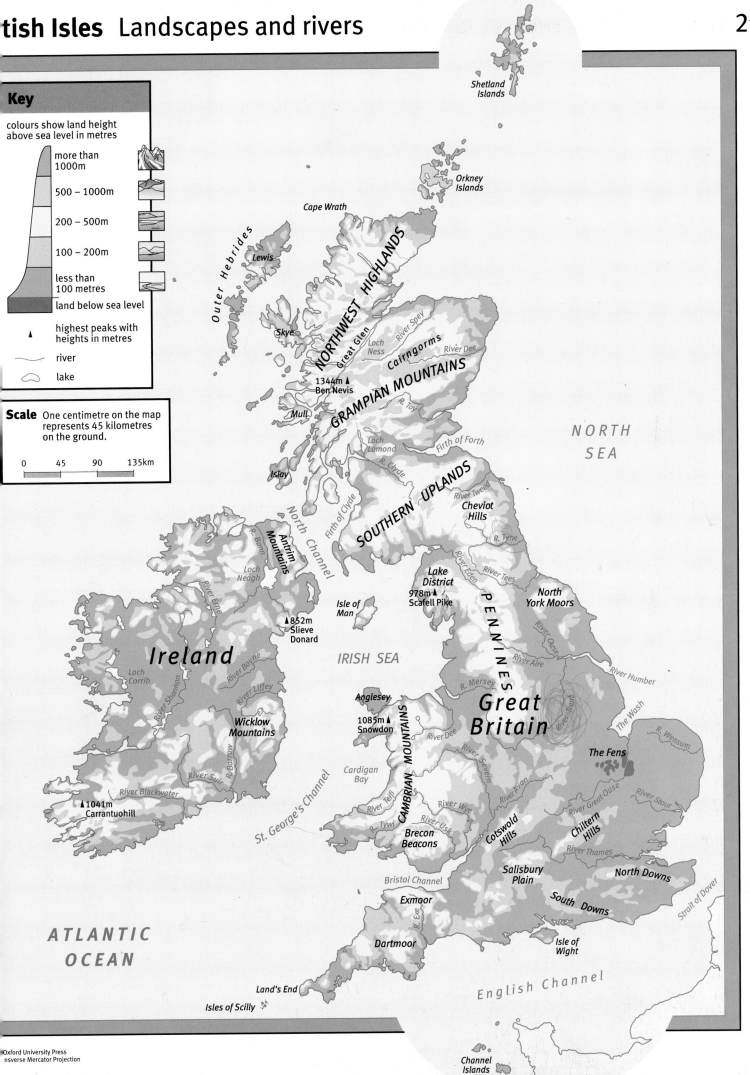

Key

colours show land height
above sea level in metres

more than
1000m

500 – 1000m

200 – 500m

100 – 200m

less than
100 metres

land below sea level

▲ highest peaks with
heights in metres

river

lake

Scale One centimetre on the map
represents 45 kilometres
on the ground.

0   45   90   135km

Shetland
Islands

Orkney
Islands

Cape Wrath

Outer Hebrides

Lewis

Skye

NORTHWEST HIGHLANDS

Great Glen

Loch
Ness

River Spey

Cairngorms

River Dee

1344m ▲
Ben Nevis

GRAMPIAN MOUNTAINS

R. Tay

Mull

Loch
Lomond

Firth of Forth

NORTH
SEA

Islay

R. Clyde

Firth of Clyde

SOUTHERN UPLANDS

River Tweed

Cheviot
Hills

R. Tyne

North Channel

Antrim
Mountains

Loch
Neagh

River Erne

River Bann

Isle of
Man

Lake
District
978m ▲
Scafell Pike

River Eden

River Tees

North
York Moors

PENNINES

▲852m
Slieve
Donard

IRISH SEA

River Ouse

Ireland

Loch
Corrib

River Boyne

River Liffey

River Shannon

Wicklow
Mountains

River Aire

River Humber

Anglesey

1085m ▲
Snowdon

CAMBRIAN MOUNTAINS

River Dee

R. Mersey

Great
Britain

River Trent

The Wash

R. Wensum

R. Barrow

R. Suir

River Blackwater

▲1041m
Carrantuohill

Cardigan
Bay

River Teifi

River Wye

River Severn

River Avon

The Fens

River Great Ouse

River Stour

St. George's Channel

R. Tywi

River Usk

Brecon
Beacons

Cotswold
Hills

Chiltern
Hills

River Thames

ATLANTIC
OCEAN

Bristol Channel

Exmoor

R. Exe

Salisbury
Plain

South Downs

North Downs

Strait of Dover

Dartmoor

Isle of
Wight

Land's End

English Channel

Isles of Scilly

Oxford University Press
nsverse Mercator Projection

Channel
Islands

Winter

| Jan. | Feb. | Mar. | Apr. | May | June | July | Aug. | Sep. | Oct. | Nov. | Dec. |

average temperature

16°C
14°C
12°C
10°C
8°C
6°C
4°C
2°C
0°C
-2°C

Summer

| Jan. | Feb. | Mar. | Apr. | May | June | July | Aug. | Sep. | Oct. | Nov. | Dec. |

average temperature

16°C
14°C
12°C
10°C
8°C
6°C
4°C
2°C
0°C
-2°C

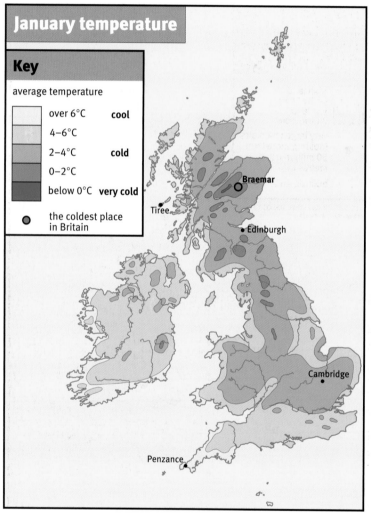

January temperature

Key

average temperature

- over 6°C **cool**
- 4–6°C
- 2–4°C **cold**
- 0–2°C
- below 0°C **very cold**
- ● the coldest place in Britain

Braemar
Tiree
Edinburgh
Cambridge
Penzance

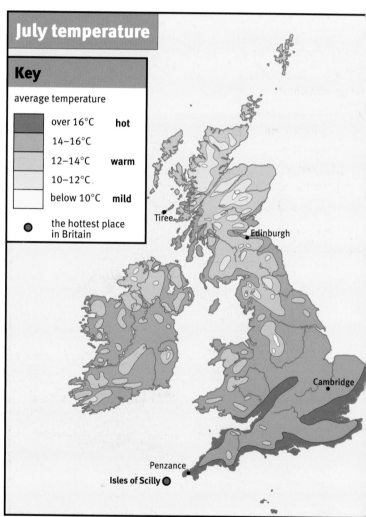

July temperature

Key

average temperature

- over 16°C **hot**
- 14–16°C
- 12–14°C **warm**
- 10–12°C
- below 10°C **mild**
- ● the hottest place in Britain

Tiree
Edinburgh
Cambridge
Penzance
Isles of Scilly ●

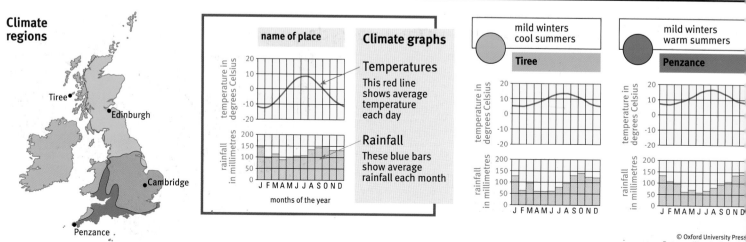

Climate regions

Tiree
Edinburgh
Cambridge
Penzance

name of place

Climate graphs

temperature in degrees Celsius

20
10
0
-10
-20

Temperatures
This red line shows average temperature each day

rainfall in millimetres

200
150
100
50
0
J F M A M J J A S O N D

months of the year

Rainfall
These blue bars show average rainfall each month

mild winters cool summers

Tiree

temperature in degrees Celsius

20
10
0
-10
-20

rainfall in millimetres

200
150
100
50
0
J F M A M J J A S O N D

mild winters warm summers

Penzance

temperature in degrees Celsius

20
10
0
-10
-20

rainfall in millimetres

200
150
100
50
0
J F M A M J J A S O N D

© Oxford University Press
Transverse Mercator Projection

Wet winds from the west rise and cool to give rain and snow. Mountains are the wettest places in the British Isles.

West is wetter

East is drier

average rainfall

- 2400mm — very wet
- 2200mm — quite wet
- 2000mm
- 1800mm
- 1600mm — wet
- 1400mm
- 1200mm
- 1000mm — quite dry
- 800mm
- 600mm — very dry
- 400mm
- 200mm

| . | Feb. | Mar. | Apr. | May | June | July | Aug. | Sep. | Oct. | Nov. | Dec. |

Annual rainfall

Key

average rainfall in a year

- more than 2400mm
- 1500–2400mm
- 800–1500mm
- 600–800mm
- less than 600mm
- ● the wettest place in Britain
- ○ the driest place in Britain

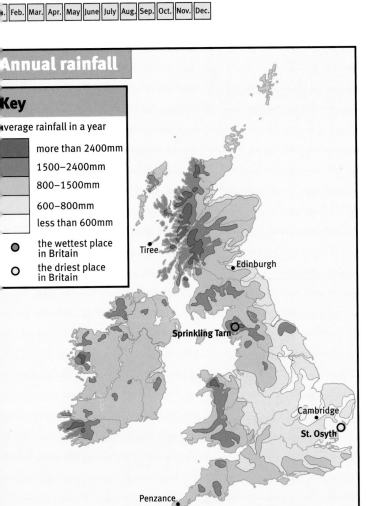

Tiree

Edinburgh

Sprinkling Tarn

Cambridge

St. Osyth

Penzance

Water supply

Key

- high land (above 200m)
- low land (below 200m)
- ● very large reservoirs (holding more than 50 million cubic metres of water*)
- built-up area

*You could fit more than 100 000 houses into these reservoirs

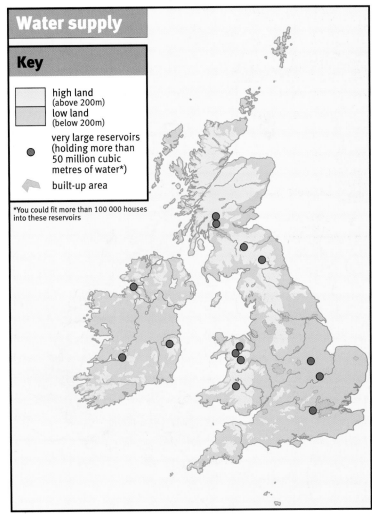

cold winters cool summers

Edinburgh

cool winters warm summers

Cambridge

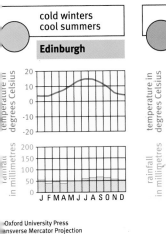

temperature in degrees Celsius

20
10
0
-10
-20

rainfall in millimetres

200
150
100
50

J F M A M J J A S O N D

temperature in degrees Celsius

20
10
0
-10
-20

rainfall in millimetres

200
150
100
50

J F M A M J J A S O N D

Oxford University Press
Transverse Mercator Projection

The water cycle

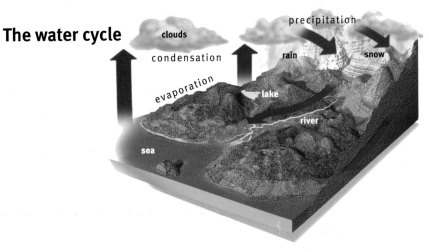

precipitation

clouds

condensation

rain

snow

evaporation

lake

river

sea

Cities and towns
People live in settlements of different sizes

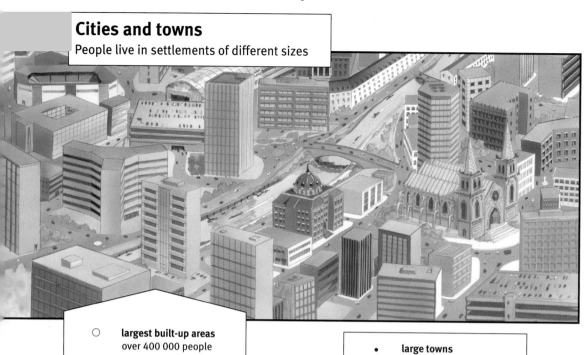

○ **largest built-up areas**
over 400 000 people

● **large towns**
25 000 – 100 000

◉ **largest towns**
100 000 – 400 000 people

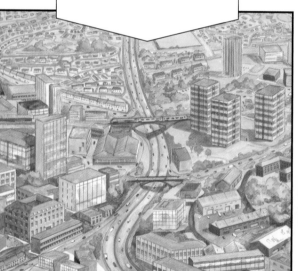

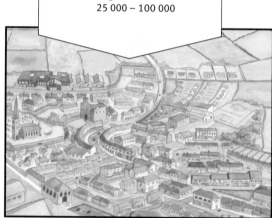

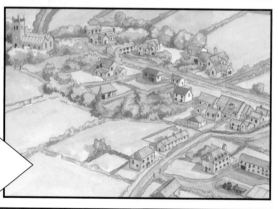

small towns and villages
under 25 000 people
(not shown on the map)

Where people live
If there were 100 people in the United Kingdom, this is where they would live:

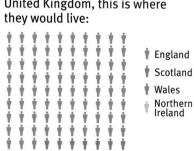

♦ England
♦ Scotland
♦ Wales
♦ Northern Ireland

Population pyramid
If there were 100 people in the United Kingdom, this is how old they would be:

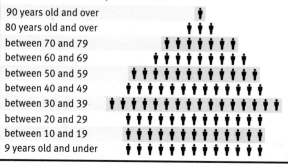

90 years old and over	
80 years old and over	
between 70 and 79	
between 60 and 69	
between 50 and 59	
between 40 and 49	
between 30 and 39	
between 20 and 29	
between 10 and 19	
9 years old and under	

Population density
The number of people that live in an area

very crowded
over 250 people living in a square kilometre

quite crowded
50 – 250 people living in a square kilometre

quite empty
under 50 people living in a square kilometre

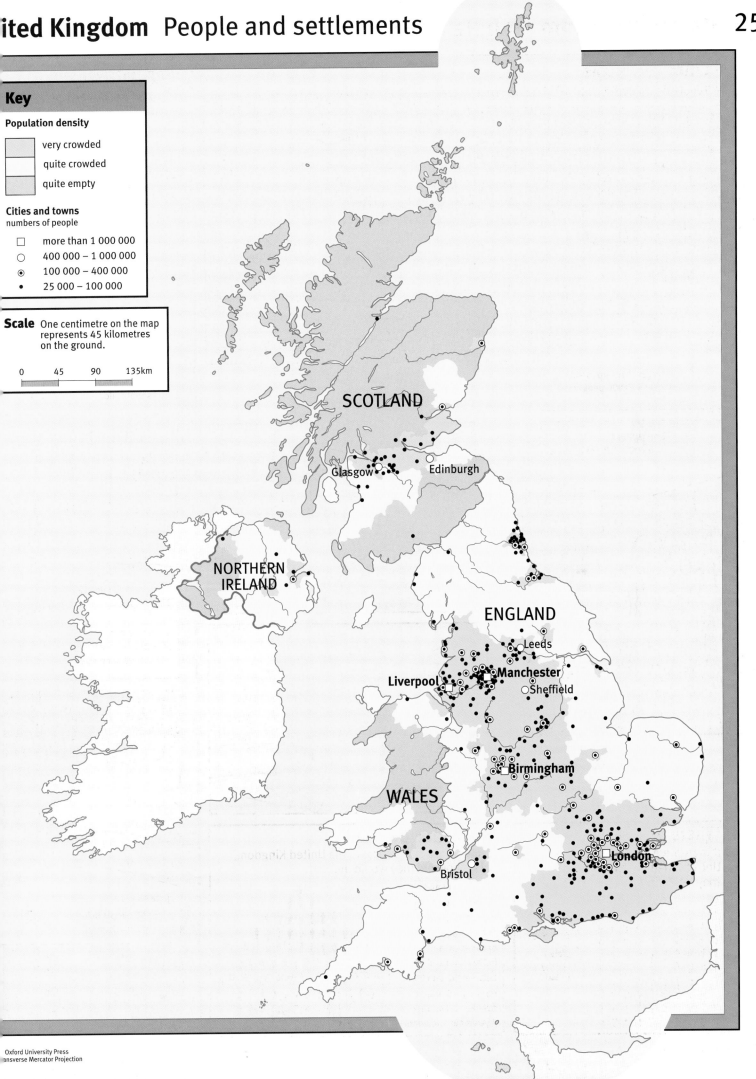

Key

Population density

	very crowded
	quite crowded
	quite empty

Cities and towns
numbers of people

☐	more than 1 000 000
○	400 000 – 1 000 000
◉	100 000 – 400 000
•	25 000 – 100 000

Scale
One centimetre on the map represents 45 kilometres on the ground.

0   45   90   135km

SCOTLAND

Glasgow   Edinburgh

NORTHERN
IRELAND

ENGLAND

Leeds

Liverpool   Manchester

Sheffield

Birmingham

WALES

Bristol   London

Oxford University Press
Transverse Mercator Projection

United Kingdom Air and sea connections

Key

- major hub airport
- major airport
- car ferry route
- major car ferry port
- major sea port
- built-up area

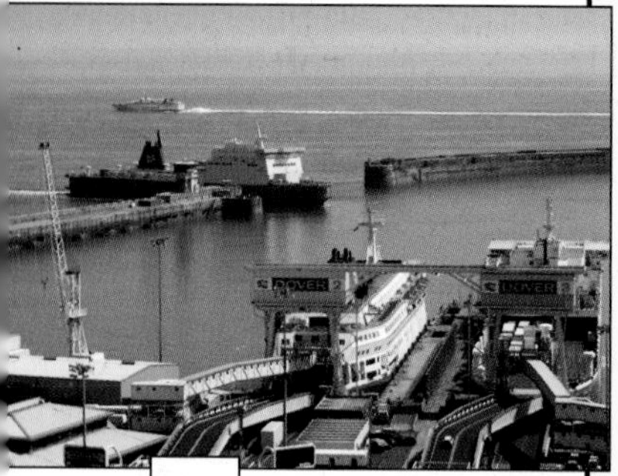

major car ferry port, Dover

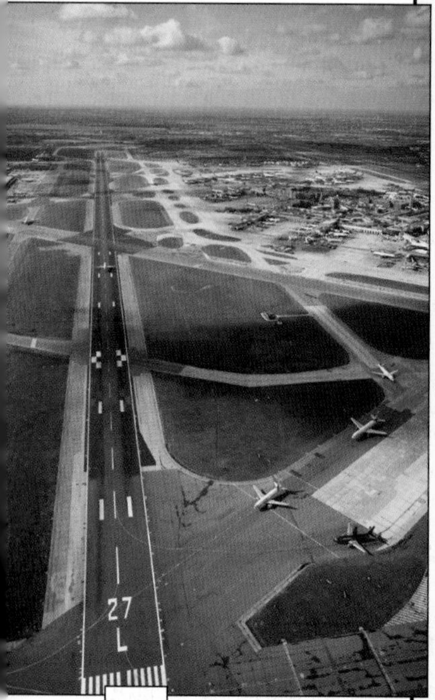

major hub airport, London Heathrow

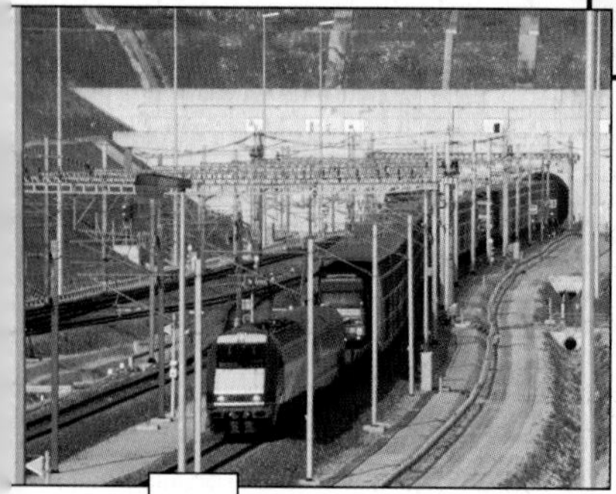

Channel Tunnel terminal, Ashford

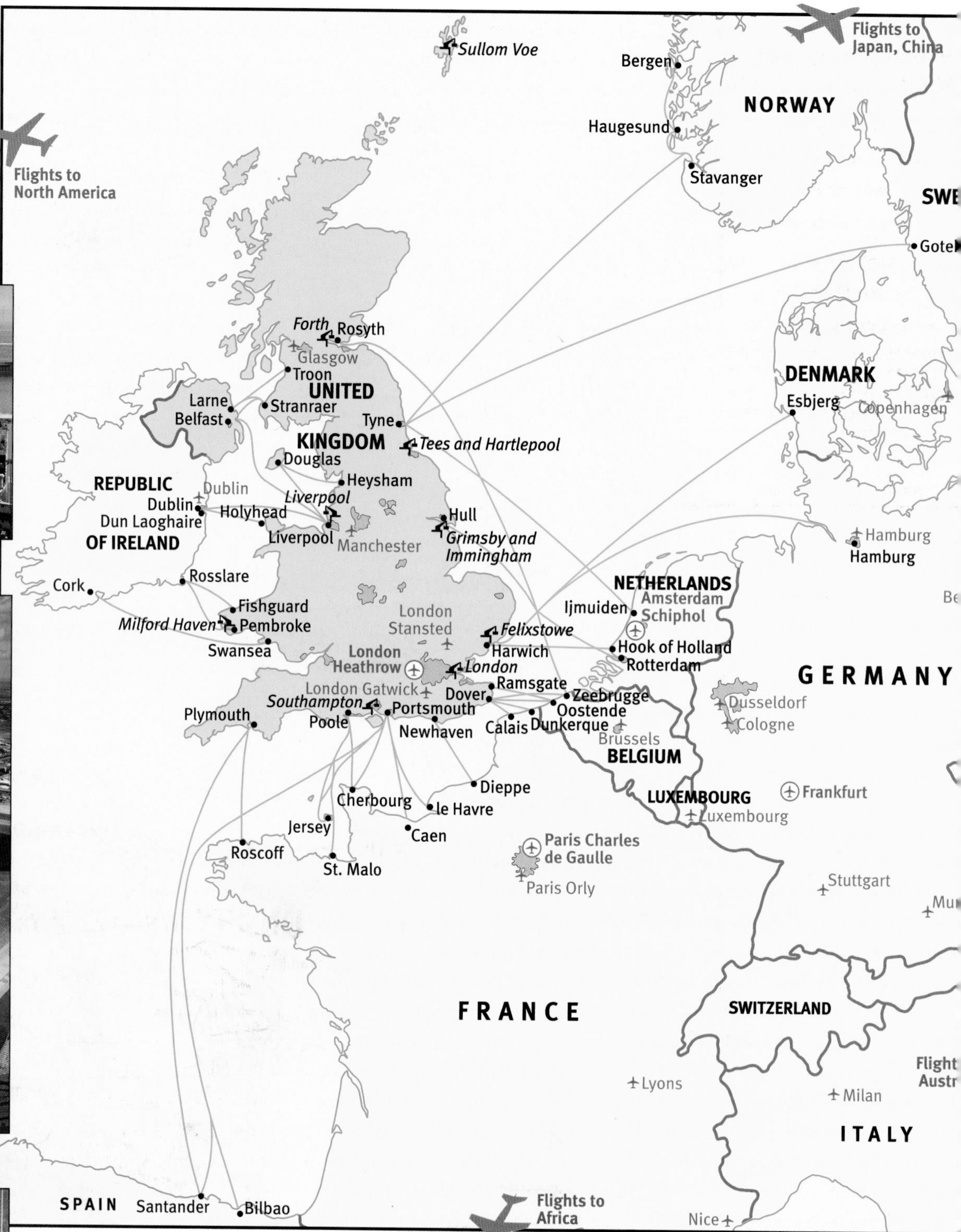
Sullom Voe
Bergen
Haugesund
Stavanger
NORWAY
Flights to Japan, China
Flights to North America
SWE
Gote
Forth
Rosyth
Glasgow
Troon
UNITED KINGDOM
Larne
Belfast
Stranraer
Tyne
Tees and Hartlepool
Douglas
Heysham
REPUBLIC OF IRELAND
Dublin
Dublin
Dun Laoghaire
Holyhead
Liverpool
Liverpool
Manchester
Hull
Grimsby and Immingham
DENMARK
Esbjerg
Copenhagen
Hamburg
Hamburg
Cork
Rosslare
Fishguard
Milford Haven
Pembroke
Swansea
London Stansted
Felixstowe
Harwich
London Heathrow
London
NETHERLANDS
Ijmuiden
Amsterdam Schiphol
Hook of Holland
Rotterdam
GERMANY
Be
London Gatwick
Ramsgate
Dover
Zeebrugge
Oostende
Southampton
Portsmouth
Plymouth
Poole
Newhaven
Calais
Dunkerque
Brussels
BELGIUM
Dusseldorf
Cologne
Dieppe
Cherbourg
le Havre
LUXEMBOURG
Luxembourg
Frankfurt
Jersey
Caen
Roscoff
St. Malo
Paris Charles de Gaulle
Paris Orly
Stuttgart
Mu
FRANCE
SWITZERLAND
Lyons
Milan
Flight Austr
ITALY
SPAIN
Santander
Bilbao
Flights to Africa
Nice

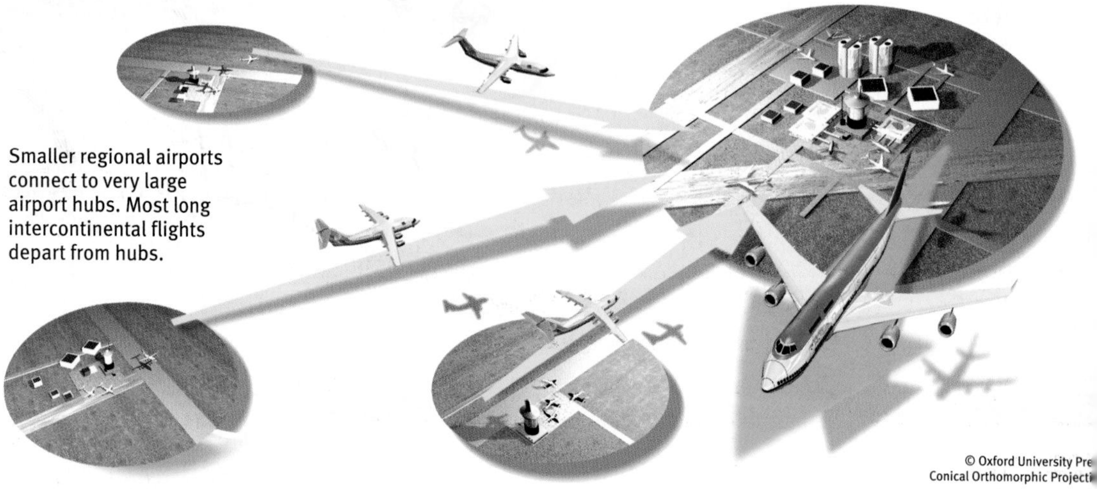

Smaller regional airports connect to very large airport hubs. Most long intercontinental flights depart from hubs.

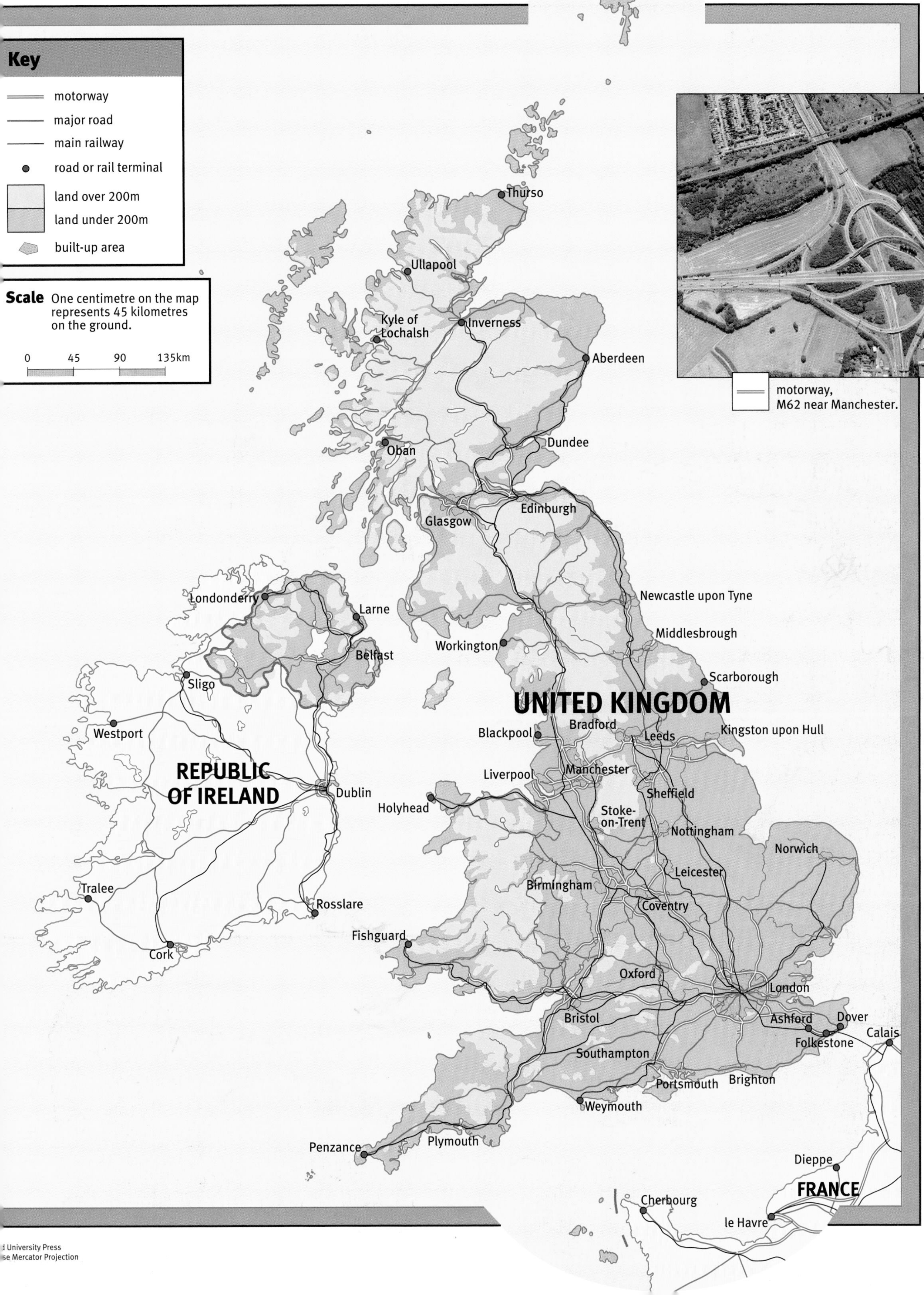

Key

═══	motorway
───	major road
───	main railway
●	road or rail terminal
	land over 200m
	land under 200m
	built-up area

Scale One centimetre on the map represents 45 kilometres on the ground.

0   45   90   135km

motorway,
M62 near Manchester.

Thurso

Ullapool

Kyle of
Lochalsh

Inverness

Aberdeen

Oban

Dundee

Edinburgh

Glasgow

Newcastle upon Tyne

Middlesbrough

Londonderry

Larne

Belfast

Workington

Scarborough

Sligo

UNITED KINGDOM

Westport

Bradford

Blackpool

Leeds

Kingston upon Hull

**REPUBLIC
OF IRELAND**

Liverpool

Manchester

Dublin

Holyhead

Sheffield

Stoke-
on-Trent

Nottingham

Norwich

Tralee

Leicester

Rosslare

Birmingham

Coventry

Fishguard

Cork

Oxford

London

Bristol

Ashford

Dover

Calais

Folkestone

Southampton

Portsmouth

Brighton

Weymouth

Plymouth

Penzance

Dieppe

FRANCE

Cherbourg

le Havre

d University Press
se Mercator Projection

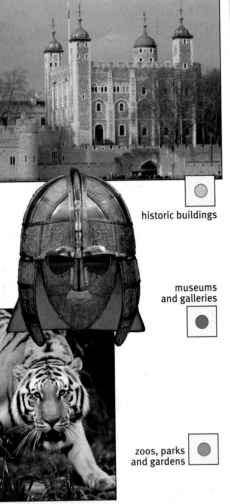

Top UK tourist attractions

Key

Symbol **colour** shows the type of tourist attraction

- historic buildings
- museums and galleries
- zoos, parks and gardens
- theme parks and piers

Symbol **size** shows how popular the attraction is

- ○ over 4 million visitors each year
- ○ 2–4 million visitors each year
- ○ 1–2 million visitors each year
- built-up area

historic buildings

museums and galleries

zoos, parks and gardens

theme parks and piers

Drumpellier Country Park
Edinburgh Castle
Kelvingrove Art Gallery and Museum
Strathclyde Country Park

Windermere Lake Cruises
Flamingoland Theme Park and Zoo, Kirby Misperton
York Minster
Blackpool Pleasure Beach
Pleasureland Theme Park Southport
Upper Derwent Reservoirs
Alton Towers
Chester Zoo
Drayton Manor Family Theme Park, Tamworth
Pleasure Beach, Great Yarmouth
Fairlands Valley Park
Ashton Court Estate
Legoland, Windsor
Kew Gardens
Canterbury Cathedral
Eastbourne Pier
Eden Project

Inner London

Madam Tussaud's
British Museum
National Gallery
National Portrait Gallery
London Eye
Science Museum
Tower of London
Natural History Museum
Tate Modern
Tate Britain
Victoria & Albert Museum
Westminster Abbey

Holidays abroad

Holidays in the UK and abroad

Number of holidays taken by people who live in the UK

millions

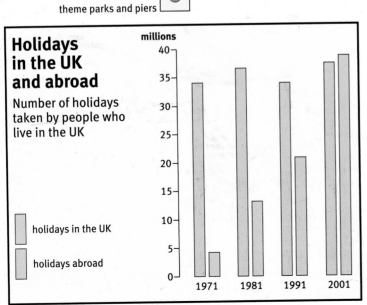

| | holidays in the UK |
| | holidays abroad |

1971   1981   1991   2001

Canada
USA

each symbol stands for 1 million British tourists

© Oxford University Pre

Key

- National Parks
- areas of outstanding scenery and beauty
- protected coast
- ✳ World Heritage site
- built-up area

Scale One centimetre on the map represents 45 kilometres on the ground.

0   45   90   135km

Shetland

Hoy and West Mainland

✳ The Heart of Neolithic Orkney

South Lewis, Harris, and North Uist

Kyle of Tongue

Assynt Coigach

✳ St. Kilda

Wester Ross

The Cuillin Hills

Knoydart

Cairngorms

Aberdeen

Ben Nevis and Glen Coe

Loch Rannoch and Glen Lyon

Loch na Keal, Isle of Mull

Loch Lomond and The Trossachs

Knapdale

Old and New Towns of Edinburgh

Jura

Edinburgh

North Arran

Glasgow

New Lanark ✳

Upper Tweeddale

Giant's Causeway ✳

Antrim Coast and Glens

Northumberland

Sperrin

Belfast

Hadrian's Wall ✳

Newcastle upon Tyne

Strangford Lough

North Pennines

✳ Durham Cathedral/Castle

Mourne

Lake District

North York Moors

Yorkshire Dales

Nidderdale

✳ Fountain's Abbey/Studley Royal Park

Forest of Bowland

Saltaire ✳   Leeds

Liverpool - Maritime Mercantile City ✳

Manchester

Lincolnshire Wolds

Anglesey

Liverpool

Sheffield

Norfolk Coast

Peak District

Castles/Town Walls of King Edward ✳

Clwydian Range

Derwent Valley Mills

Lleyn

Stoke-on-Trent

Nottingham

Snowdonia

The Broads

Ironbridge Gorge ✳

Shropshire Hills

Coventry

Suffolk Coast and Heaths

Birmingham

Pembrokeshire Coast

Brecon Beacons

Wye Valley

Cotswolds

Blenheim Palace ✳

London

Blaenavon ✳

Oxford

Chilterns

Tower of London ✳

Gower

Bristol

Kew Gardens ✳

✳ Maritime Greenwich

Cardiff

✳ Bath

North Wessex Downs

Westminster Palace/Abbey ✳

✳ Canterbury Cathedral

Exmoor

Cranborne Chase

✳ Stonehenge/Avebury

South Downs

Kent Downs

High Weald

Blackdown Hills

Dorset

Isle of Wight

Dartmoor

Dorset and East Devon Coast

New Forest

Cornwall

Isles of Scilly

Tamar Valley

National Park Snowdonia

Area of outstanding scenery and beauty The Cotswolds

Protected coast Pembrokeshire coast

World Heritage site Ironbridge

Oxford University Press
Transverse Mercator Projection

A B C D E F

3

Arctic Circle

GREENLAND
SEA

Prime Meridian

Iceland
▲1491m
Mount
Hekla

N

60°N

Lofoten
Islands

Kola
Peninsula

Lappland

WHITE
SEA

River Pechora

URAL MOUNTAINS

Faroe
Islands

Shetland
Islands

Scandinavia

Gulf of Bothnia

River North Dvina

Lake
Onega

ATLANTIC
OCEAN

Orkney
Islands

Lake
Ladoga

2

NORTH
SEA

Lake
Vänern

Gotland

Lake
Peipus

Rybinsk
Reservoir

River Volga

Ireland

Great
Britain

Lake
Vättern

BALTIC SEA

Central
Russian
Uplands

R. Thames

Jylland

Bornholm

English Channel

Friesian Islands

River Elbe

North European Plain

River Oder

River Vistula

River Dniepr

River Don

Cape
Finisterre

Bay of
Biscay

River Seine

R. Rhine

River Donets

Cantabrian Mts.

River Loire

Jura

R. Dniester

CARPATHIANS

SEA OF
AZOV

CASPIAN
SEA

Massif
Central

4807m
Mont Blanc

ALPS

River Po

Hungarian
Basin

River Duero

River Ebro

Pyrénées

R. Rhône

2548m▲

CAUCASUS

R. Tagus

River Danube

BLACK SEA

5642m
Mt. Elbrus

Cape
St. Vincent

Corsica

APPENNINES

Dinaric Alps

ADRIATIC SEA

5123m
Mt. Ararat

Balearic Islands

Menorca

Sardinia

Pindus Mountains

Anatolian
Plateau

Lake
Van

40°E

Ibiza

Mallorca

TYRRHENIAN
SEA

2917m
▲Mt. Olympus

AEGEAN SEA

Taurus Mountains

Str. of
Gibraltar

M E D I T E R R A N E A N

Mt. Etna
3323m▲

Sicily

IONIAN
SEA

Peloponnese

Rhodes

Malta

Crete

Cyprus

S E A

1

Key

land height in metres
above sea level

more than
2000m

1000 – 2000m

500 – 1000m

200 – 500m

less than
200 metres

land below
sea level

▲ highest peaks with
heights in metres

lake

river

Scale One centimetre on the map
represents 240 kilometres
on the ground.

0   240   480   720km

Tropic of Cancer

20°E

C D

A B C D E F

20°W 0° 20°E 40°E 60°E

Arctic Circle

3

50°N

ICELAND
■ Reykjavik

N

60°N

ATLANTIC

OCEAN

NORTH
SEA

**RUSSIAN
FEDERATION
(RUSSIA)**

FINLAND

■ Helsinki   • St. Petersburg

■ Stockholm
■ Oslo

• Nizhniy-
Novgorod

ESTONIA
■ Tallinn

Baltic Sea

2

60°N

Belfast   • Edinburgh

• Göteborg

LATVIA
■ Riga

■ Moscow

**REPUBLIC
OF IRELAND**
■ Dublin

UNITED

KINGDOM
• Manchester

DENMARK
Copenhagen ■

LITHUANIA

KALININGRAD
(Russia)

■ Vilnius

• Birmingham

NETHERLANDS

• Hamburg

■ Minsk

POLAND

BELARUS

• London

Rotterdam •
■ Amsterdam

■ Berlin

BELGIUM

GERMANY
• Düsseldorf

■ Warsaw

LUXEMBOURG

■ Brussels

■ Kiev

• Kharkov

• Volgograd

■ Paris

■ Luxembourg

■ Prague

U K R A I N E

• Krakow
CZECH REP.

• Donets'k

• Rostov-on-Don

FRANCE

• Munich

LIECHTENSTEIN

SLOVAKIA
■ Vienna ■ Bratislava

MOLDOVA

• Bern

AUSTRIA

■ Budapest

■ Chisinau

• Odessa

• Bordeaux

• Lyons

SWITZERLAND

■ Ljubljana
HUNGARY

GEORGIA

40°N

• Milan

SLOVENIA

■ Zagreb

ROMANIA

■ Tbilisi

• Oporto

PORTUGAL

ANDORRA

MONACO

CROATIA

BOSNIA–
HERZEGOVINA

■ Belgrade

■ Bucharest

B L A C K   S E A

SPAIN

• Marseilles

I T A L Y

SAN
MARINO

■ Sarajevo

**SERBIA AND
MONTENEGRO**

BULGARIA

■ Lisbon

■ Madrid

• Barcelona

■ Rome

■ Sofia

• Istanbul

T U R K E Y

• Valencia

• Naples

■ Tiranë

■ Skopje
FYRO
MACEDONIA

• Seville

• Gibraltar
(UK)

ALBANIA

■ Ankara

Ceuta
(Sp.)

Melilla
(Sp.)

GREECE

• Izmir

• Adana

MOROCCO

M E D I T E R R A N E A N

■ Athens

S E A

■ Valletta
MALTA

■ Nicosia

CYPRUS

SYRIA

IRAQ

1

TUNISIA

LEBANON
ISRAEL

JORDAN

40°E

LIBYA

EGYPT

SAUDI
ARABIA

0°

20°E

Tropic of Cancer

B

C

D

Key

colours show
countries

ITALY   country names are
labelled like this

■   capital cities

•   other important cities

Locator

Key

———	country boundary
- - -	disputed boundary
———	motorway or main road
———	railway
✈	main airport
~~	river
⌒	lake

towns and cities

■ capital cities

○ largest towns

• other large towns

land height

above sea level in metres

more than 5000m

2000 – 5000m

1000 – 2000m

500 – 1000m

200 – 500m

less than 200 metres

land below sea level

▲ highest peaks with heights in metres

Scale One centimetre on the map represents 150 kilometres on the ground.

0   150   300   450km

ATLANTIC OCEAN

NORTH SEA

20°W   60°N   10°W   0°   10°E

Shetland Islands

Outer Hebrides

Orkney Islands

Inverness

1344m Ben Nevis

Aberdeen

Glasgow   Dundee   Edinburgh

Galway   Belfast

REPUBLIC OF IRELAND   Dublin   Manchester   UNITED

Cork   Birmingham   KINGDOM

Cardiff

London

English Channel

Brest

NORWAY   Bergen   Oslo

Lake Väne

Göte

DENMARK   Copenhagen

Frisian Is.

NETHERLANDS   Hamburg   Szcz

The Hague   Amsterdam   R. Elbe   Bore

Rotterdam   Hannover

BELGIUM   Düsseldorf

Brussels   GERMANY

LUXEMBOURG

Paris   Luxembourg   Prague   CZ

Strasbourg   Nürnberg   R. Danube

R. Seine   R. Rhine

Nantes   FRANCE   Munich   AUSTR

R. Loire   Bern   LIECHTENSTEIN

MASSIF CENTRAL   SWITZERLAND   Ljubljana

Bay of Biscay   4807m   Milan   Za

Bordeaux   R. Garonne   Mont Blanc   A L P S

A Coruña   Toulouse   R. Rhône   Turin   Verona

Cape Finisterre   PYRÉNÉES   MONACO   Florence   SAN MARINO

Oporto   Cantabrian Mts.   Bilbao   ANDORRA   Marseilles   ITALY

R. Douro   Zaragoza   Corsica (France)   APPENNINES

PORTUGAL   R. Duero   Ajaccio

SPAIN   R. Ebro   Rome

Lisbon   Madrid   Barcelona   Naples

R. Tagus   Valencia   Balearic Islands   Sássari

Faro   Seville   R. Guadalquivir   Menorca   Sardinia (Italy)   TYRRHENIAN

Cape St. Vincent   Cádiz   Ibiza   Mallorca   Cágliari   SEA

Tangier   Gibraltar (UK)   M E D I T E R R A N E A N   Réggio Calab

Ceuta (Sp.)   Melilla (Sp.)   Oran   Algiers   Palermo

Rabat   Annaba   Mt Etna 3323m   Sicily

Casablanca   Fès   Constantine   Tunis   Vall

TUNISIA   MALT

MOROCCO   Sfax

A T L A S   M O U N T A I N S

Béchar

30°N   Touggourt

A L G E R I A   Tripoli

Misratah

40°N   50°N

0°   10°E   30°N

LIBYA

© Oxford University Press

Conical Orthomorphic Projection

FINLAND
Helsinki
Tallinn
Stockholm
ESTONIA
Gotland
LATVIA
Riga
LITHUANIA
Kaliningrad RUSSIA
Vilnius
dansk
North European Plain
BELARUS
Minsk
POLAND
Warsaw
Wroclaw
Krakow
L'viv
EP.
SLOVAKIA
Bratislava
enna
Budapest
HUNGARY
CARPATHIANS
R. Dniester
MOLDOVA
Chisinau
ROMANIA
BOSNIA-
HERZEGOVINA
Belgrade
Bucharest
Sarajevo
R. Danube
ric Alps
SERBIA
AND
MONTENEGRO
Sofia
BULGARIA
Tiranē
Skopje
FYRO
MACEDONIA
ALBANIA
Thessaloníki
Mt. Olympus
2917m
Taranto
GREECE
Pindus Mts.
IONIAN
SEA
Athens
Peloponnese
AEGEAN
SEA
Rhodes
Iraklión
Crete
SEA
Benghazi

Gulf of Bothnia
20°E
30°E
Lake Onega
40°E
60°N
50°E
Lake Ladoga
Vologda
St. Petersburg
Rybinsk Reservoir
R. Volga
Lake Peipus
Nizhniy-Novgorod
Kazan'
Baltic Sea
G. of Riga
Moscow
RUSSIAN FEDERATION (RUSSIA)
Samara
R. Volga
50°E
R. Dnieper
Kiev
Kharkov
Volgograd
UKRAINE
R. Don
Dnipropetrovsk
Donets'k
Rostov-on-Don
Odessa
SEA OF AZOV
Crimea
Mt. Elbrus
5642m
CAUCASUS MTS
Sevastopol
Constanta
BLACK SEA
GEORGIA
Samsun
Istanbul
Sivas
Ankara
TURKEY
Bursa
Kayseri
Izmir
Konya
Adana
Taurus Mountains
Aleppo
R. Euphrates
Nicosia
CYPRUS
SYRIA
LEBANON
Beirut
Damascus
ISRAEL
Amman
Jerusalem
Dead Sea
Port Said
JORDAN
Alexandria
Cairo
30°N
El Giza
EGYPT
Sinai
SAUDI ARABIA
20°E
30°E

Bridges over the River Seine in Paris

Bridges over the
River Vltava in Prague

5

60°N · Prime Meridian · 20°W · 40°W · 80°W · 120°W · 80°W · 60°N · 160°W

North Pole

ARCTIC OCEAN

J

A

B

C

D

E

F

G

H

Bering Strait

BERING SEA

180°

Arctic Circle

0°

BARENTS SEA

Lake Ladoga

Lake Onega

River Ob

Yenisey River

Central Siberian Plateau

River Lena

Kamchatka

SEA OF OKHOTSK

Sakhalin

Kuril Islands

40°N

20°E

URAL MOUNTAINS

Siberian Lowland

Angara River

River Amur (Heilong Jiang)

160°E

40°N

River Volga

Lake Baykal

Hokkaido

4

BLACK SEA

CAUCASUS

Mt. Elbrus 5642m

Mt Ararat 5123m

Anatolian Plateau

River Irtysh

Kazakh Upland

Lake Balkhash

ALTAI MOUNTAINS

Gobi Desert

SEA OF JAPAN

Mt. Fuji 3776m

MEDITERRANEAN SEA

Caspian Sea

Aral Sea

Honshu

Dead Sea

Elburz Mts.

Qullai Garmo 7495m

TIEN SHAN

Turpan Depression −154m

Tarim Basin

YELLOW SEA

Kyushu

R. Euphrates

R. Tigris

ZAGROS MTS.

Hindu Kush

KUNLUN SHAN

Huang He

EAST CHINA SEA

Ryukyu Islands

Tropic of Cancer

20°N

The Gulf

K2 8611m

Plateau of Tibet

Chang Jiang (Yangtze R.)

20°N

Arabian Peninsula

River Indus

HIMALAYA

Mt. Everest 8848m

Taiwan

3

RED SEA

Gulf of Aden

Socotra

ARABIAN SEA

Thar Desert

River Ganges

Brahmaputra

Mouths of the Ganges

Irrawaddy R.

Salween R.

SOUTH CHINA SEA

Luzon

PACIFIC OCEAN

60°E

D e c c a n

Bay of Bengal

Mindoro

Mindanao

Laccadive Islands

Andaman Islands

ANDAMAN SEA

Gulf of Thailand

Mekong R.

4094m Mt. Kinabalu

CELEBES SEA

Nicobar Islands

Maldive Archipelago

Malay Peninsula

Borneo

Sulawesi

New Gui

0°

Equator

2

INDIAN OCEAN

S u m a t r a

JAVA SEA

Timor

ARAFURA SEA

Java

Bali

TIMOR SEA

20°S

80°E

N

120°E

Tropic of Capricorn

100°E

20°S

Key

land height in metres
above sea level

more than
5000m

2000 – 5000m

1000 – 2000m

500 – 1000m

200 – 500m

less than
200 metres

land below sea level

▲ highest peaks with
heights in metres

lake

river

Scale One centimetre on the map
represents 550 kilometres
on the ground.

0   550   1100   1650km

Zenithal Equal Area Projection

B · C · D · E

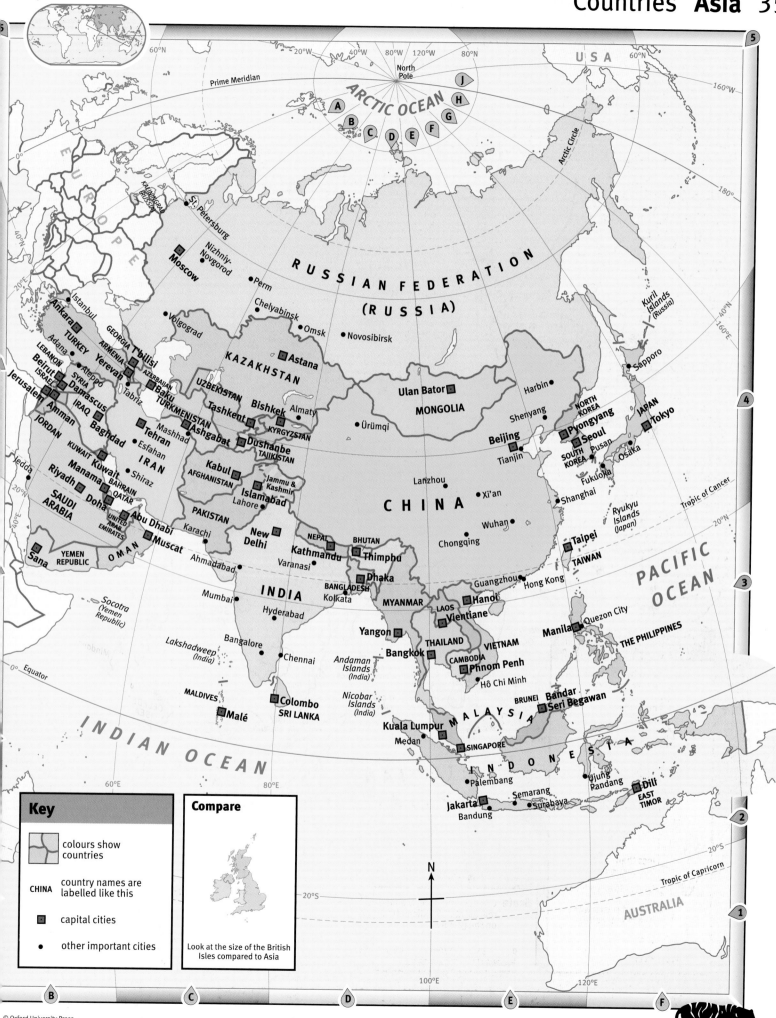

Key

colours show countries

CHINA country names are labelled like this

▢ capital cities

• other important cities

Compare

Look at the size of the British Isles compared to Asia

© Oxford University Press

Key

——————	country boundary
– – – –	disputed boundary
——————	motorway or main road
——————	railway
⊕	main airport
⌒	river
⌒	lake

land height
above sea level in metres

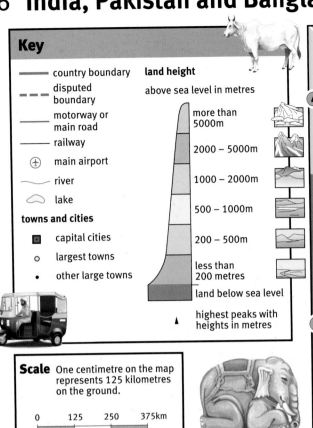

more than 5000m	
2000 – 5000m	
1000 – 2000m	
500 – 1000m	
200 – 500m	
less than 200 metres	
	land below sea level
▲	highest peaks with heights in metres

towns and cities

▣	capital cities
○	largest towns
•	other large towns

Scale One centimetre on the map represents 125 kilometres on the ground.

0   125   250   375km

Traffic jam in Rajasthan

◻ Rush hour in Jaipur

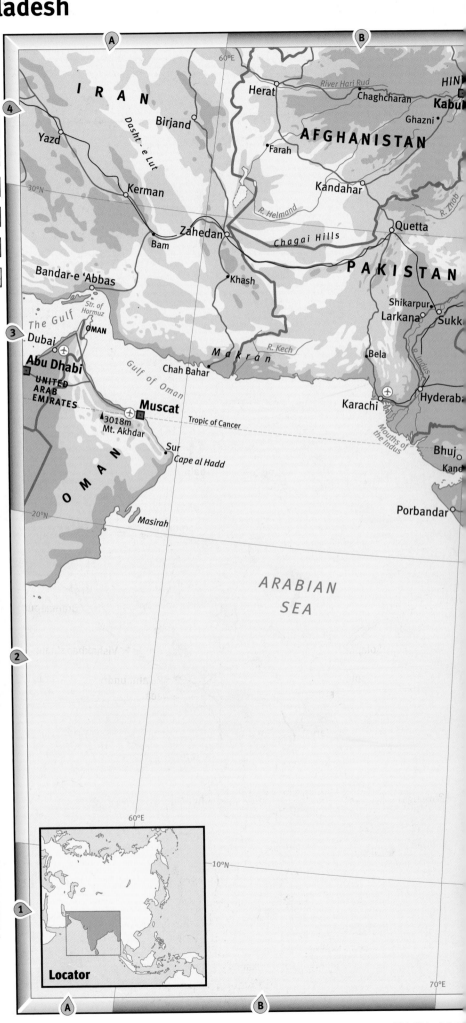

Traffic jam in Rajasthan

IRAN

Yazd
Birjand
Kerman
Zahedan
Bam
Bandar-e 'Abbas
Khash

The Gulf
Str. of Hormuz
Dubai
OMAN
Abu Dhabi
UNITED ARAB EMIRATES
Muscat
3018m Mt. Akhdar
Sur
Cape al Hadd
Masirah

OMAN

Dasht-e Lut
Herat
Chaghcharan
Kabu
Ghazni
AFGHANISTAN
Farah
Kandahar
R. Helmand
Chagai Hills
Quetta
R. Znob
PAKISTAN
Shikarpur
Larkana
Sukk
Bela
Karachi
Hyderaba
Mouths of the Indus
Bhuj
Kand
Porbandar

River Hari Rud
HIN

R. Kech
Makran
Chah Bahar
Gulf of Oman
Tropic of Cancer

ARABIAN SEA

60°E
30°N
20°N
10°N
70°E
60°E

Locator

© Oxford University Press
Conical Orthomorphic Projection

PAMIRS
TAJIKISTAN

Khorog

7690m

Gilgit

K2 (Qogir Feng, Godwin Austen) 8611m

80°E

C

D

90°E

E

awar

Srinagar JAMMU AND KASHMIR

Leh

Rutog

Islamabad

Rawalpindi

Jammu

Gujranwala

Lahore

Amritsar

Ludhiana

Chandigarh

Dehra Dun

isalabad

Multan

Ghazi Khan

River Sutlej

Bahawalpur

Bikaner

Meerut

New Delhi Delhi

Bareilly

R. Yamuna

Jaipur

Agra

odhpur

Gwalior

Kota

Jhansi

R. Banas

R. Chambal

R. Ghaghra

Lucknow

Gorakhpur

Annapurna 8091m

Kanpur

NEPAL

Kathmandu

Mount Everest 8848m

Lhaze

Thimphu BHUTAN

Darjiling

Dibrugarh

Guwahati

Nagaon

Shillong

Lhasa

Nyingchi

Yarlung Zangbo (Tsangpo R.)

CHINA

Jinsha Jiang (Yangtze R.)

Lancang Jiang (Mekong R.)

Nu Jiang (Salween R.)

30°N

4

H I M A L A Y A

River Ganges

Gomati

Muzaffarpur

Patna

Varanasi

Bhagalpur

R. Ganges

R. Son

Allahabad

BANGLADESH

Imphal

Brahmaputra R.

River Chindwin

3

Ahmadabad

Vadodara

agar

Bharuch

Surat

Nashik

Dhule

mbai

Pune

Indore

Bhopal

R. Narmada

R. Tapi

Burhanpur

Amravati

Nagpur

Nizamabad

Deccan

Murwara

Jabalpur

Bilaspur

Raipur

Chandrapur

I N D I A

Dhanbad

Asanol

Jamshedpur

Kharagpur

Sambalpur

R. Mahanadi

Cuttack

Brahmapur

Khulna

Dhaka

Kolkata

Mouths of the Ganges

Chittagong

Sittwe

Tropic of Cancer

Monywa

Mandalay

MYANMAR (BURMA)

Arakan Yoma

Irrawaddy R.

20°N

2

Hirakud Reservoir

Gandhi Sagar

Aurangabad

R. Godavari

R. Indravati

R. Godavari

Solapur

Hyderabad

Rajahmundry

Vijayawada

Vishakhapatnam

Bay of Bengal

Sandoway

Pye

Yangon

Bassein

Mouths of the Irrawaddy

Kolhapur

Bijapur

R. Bhima

Raichur

R. Krishna

Belgaum

Bellary

WESTERN GHATS

EASTERN GHATS

R. Pennel

Nellore

Bangalore

Mysore

Vellore

Chennai

Pondicherry

Andaman Islands

Port Blair

ANDAMAN SEA

Mangalore

Laccadive Islands

Calicut

Coimbatore

Salem

Tiruchchirappalli

I N D I A N   O C E A N

10°N

1

Cochin

Madurai

Jaffna

Quilon

Trivandrum

Nagercoil

SRI LANKA

Trincomalee

Batticaloa

Puttalam

Colombo

Kandy

Badulla

Galle

80°E

C

D

90°E

E

Oxford University Press

A B C D E

5

RUSSIAN FEDERATION (RUSSIA)

Pavlodar
Barnaul
Biysk
80°E
90°E
100°E
110°E
River Ob
Astana
Rubtsovsk
Angarsk Irkutsk
Lake Baykal
Chita
50°N
Semipalatinsk
Ust'-Kamenogorsk
Ulan-Ude
Borzya
Karaganda
Zyryanovsk
Manzhouli
KAZAKHSTAN
Lake Zaysan
Altay
Hovd
Uvs Nuur
Hövsgöl Nuur
Choybalsan
Ayaguz
Ulaangom
Ulan Bator
4
Lake Balkhash
Lake Alakol
Selenge River
Taldykorgan
M O N G O L I A
Almaty
Yining
Saynshand
Bishkek
Issyk Kul
Ürümqi
KYRGYZSTAN
Erenhot
TIEN SHAN
Turpan
Hami
G o b i   D e s e r t
40°N
Tarim He
Turpan Depression −154m
Hohhot
Jining
Zhangji
Kashi
Lop Nur
Yumen
Baotou
Datong
Tang
Anxi
Qilian Shan
Wuhai
Tian
Hotan He
Tarim Pendi
Yinchuan
Great Wall
Shijiazh
Huang He
K2 (Qogir Feng) 8611m
Altun Shan
Golmud
Qinghai Hu
Taiyuan
Dezhou
Jina
C
Xining
Lanzhou
Handan
JAMMU AND KASHMIR
Kunlun   Shan
H
I
N
Changzhi
A
3
R. Indus
Rutog
Plateau of Tibet
Baoji
Wei He
Xi'an
Zhengzho
Luoyang
Xuzh
Suzho
30°N
Dehra Dun
Beng
Xiangfan
Bareilly
Lhaze
Lhasa
Batang
Chengdu
Chang Jiang (Yangtze River)
Wuha
Annapurna 8091m
Yarlung Zangbo (Tsangpo R.)
Chongqing
Changde
Dongting Hu
Jingc
Poya
Lucknow
Mt. Everest 8848m
Kathmandu
Neijiang
Nancha
Kanpur
Gorakhpur
Darjiling
Thimphu
Dibrugarh
Yibin
Changsha
Zhuzho
2
Allahabad
Varanasi
Muzaffarpur
Shiliguri
BHUTAN
Zunyi
Hengyang
Ji'
Patna
Brahmaputra R.
Guiyang
Shaoyang
Murwara
Ganges
Bhagalpur
Shillong
Dali
Duyun
Gar
Jabalpur
Jamshedpur
Dhanbad
BANGLADESH
Imphal
R. Chindwin
Kunming
Guilin
Shaogua
Tropic of Cancer
Dhaka
R. Chindwin
Liuzhou
Me
Bilaspur
Khulna
Chittagong
Monywa
Nan
Wuzhou
Guangz
Raipur
Kharagpur
I N D I A
Kolkata
M Y A N M A R
Mandalay
Salween R.
Lao Cai
Nanning
Macao
Hong
20°N
Cuttack
Mouths of the Ganges
Sittwe
(B U R M A)
Kengtung
Phongsali
Pingxiang
Zhanjiang
Brahmapur
Arakan Yoma
Irrawaddy R.
Mekong R.
Song-Koi
Hanoi
Haikou
Chiang Mai
Louangphrabang
Thanh Hoa
Hainan Dao
1
Vishakhapatnam
Bay of Bengal
Pye
Vientiane
Vinh
Sanya
SOU
CHI
SE
Bassein
Pegu
Udon Thani
90°E
Yangon
Moulmein
THAILAND
Huê
Da Nang
100°E
110°E
Mouths of the Irrawaddy

B C D E

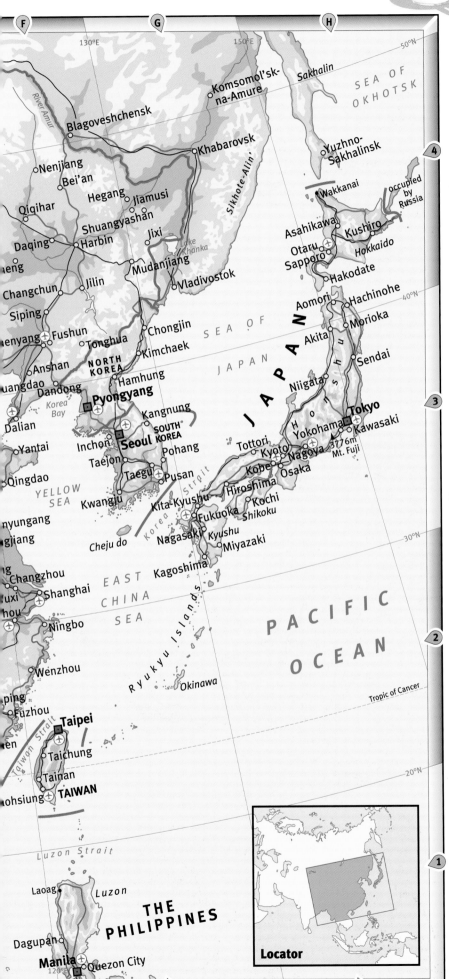

Map labels:

F 130°E **G** 150°E **H** 50°N 4

Komsomol'sk-na-Amure
Sakhalin
SEA OF OKHOTSK

River Amur
Blagoveshchensk

Nenjiang
Bei'an
Hegang
Qiqihar
Jiamusi
Daqing
Shuangyashan
...eng
Harbin
Jixi
Changchun
Jilin
Mudanjiang
Siping
...enyang
Fushun
Tonghua
Chongjin
Vladivostok
Anshan
Kimchaek
...uangdao
Dandong
Hamhung
Dalian
NORTH KOREA
Pyongyang
Korea Bay
Yantai
Kangnung
Inchon
Seoul
SOUTH KOREA
Taejon
Pohang
Qingdao
Taegu
Taegu
Pusan
YELLOW SEA
Kwangju
Cheju do

Khabarovsk
Sikhote-Alin'
Lake Khanka

Yuzhno-Sakhalinsk
Wakkanai
Asahikawa
Otaru
Sapporo
Kushiro
Hokkaido
Hakodate
Aomori
Hachinohe
Morioka
Akita
Sendai
Niigata
Tottori
Kyoto
Nagoya
Kobe
Osaka
Hiroshima
Kochi
Kita-Kyushu
Fukuoka
Shikoku
Nagasaki
Kyushu
Miyazaki
Kagoshima

SEA OF JAPAN
J A P A N
Honshu
Tokyo
Yokohama
Kawasaki
3376m Mt. Fuji

occupied by Russia

40°N
30°N

Korea Strait

nyungang
...gjiang
...ng
Changzhou
Shanghai
...uxi
...hou
Ningbo
Wenzhou
...ping
Fuzhou
Luzon Strait

EAST CHINA SEA
Ryukyu Islands
Okinawa

PACIFIC OCEAN

Tropic of Cancer
20°N

Taipei
Taichung
Tainan
...ohsiung
TAIWAN
Taiwan Strait

Laoag
Luzon
THE PHILIPPINES
Dagupan
Manila
Quezon City
120°E

F **G**

Mt. Fuji is Japan's highest peak

Shopping in Shanghai, China

Key

——	country boundary
- - -	disputed boundary
——	motorway or main road
——	railway
✈	main airport
∿	river
⬭	lake

towns and cities

▣	capital cities
○	largest towns
•	other large towns

land height
above sea level in metres

more than 5000m
2000 – 5000m
1000 – 2000m
500 – 1000m
200 – 500m
less than 200 metres
land below sea level

▲ highest peaks with heights in metres

Scale One centimetre on the map represents 180 kilometres on the ground.

0   180   360   540km

Locator

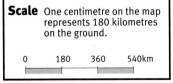

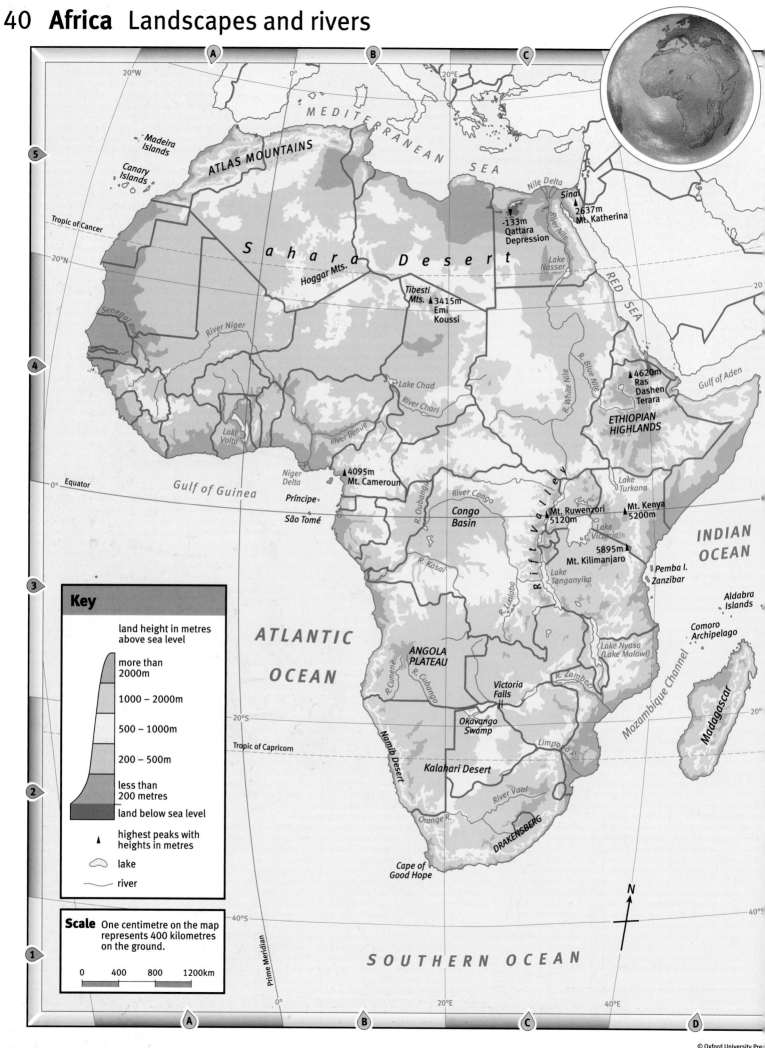

20°W   0°   20°E

MEDITERRANEAN   SEA

Madeira
Islands

ATLAS MOUNTAINS

Canary
Islands

Tropic of Cancer

Nile Delta

Sinai
2637m
Mt. Katherina

-133m
Qattara
Depression

River Nile

20°N

S a h a r a   D e s e r t

Lake
Nasser

RED SEA

20

Hoggar Mts.

Senegal River

Tibesti
Mts. ▲3415m
Emi
Koussi

River Niger

▲4620m
Ras
Dashen
Terara

R. Blue Nile

R. White Nile

Gulf of Aden

Lake Chad

ETHIOPIAN
HIGHLANDS

River Chari

Lake
Volta

River Benue

Niger
Delta

▲4095m
Mt. Cameroun

0° Equator

Gulf of Guinea

Príncipe

São Tomé

R. Oubangui

River Congo

Congo
Basin

Lake
Turkana

Mt. Kenya
5200m

Mt. Ruwenzori
5120m

Lake
Victoria

Rift Valley

INDIAN
OCEAN

R. Kasai

5895m▲
Mt. Kilimanjaro

R. Lualaba

Lake
Tanganyika

Pemba I.
Zanzibar

Aldabra
Islands

ATLANTIC

OCEAN

Comoro
Archipelago

Key

land height in metres
above sea level

more than
2000m

1000 – 2000m

500 – 1000m

200 – 500m

less than
200 metres

land below sea level

▲ highest peaks with
heights in metres

lake

river

ANGOLA
PLATEAU

R. Cunene

R. Cubango

Lake Nyasa
(Lake Malawi)

R. Zambezi

Victoria
Falls

Madagascar

Mozambique Channel

20°S

Tropic of Capricorn

Okavango
Swamp

Limpopo R.

Namib Desert

Kalahari Desert

River Vaal

Orange R.

DRAKENSBERG

Cape of
Good Hope

N

Scale One centimetre on the map
represents 400 kilometres
on the ground.

0   400   800   1200km

40°S

SOUTHERN   OCEAN

Prime Meridian

0°   20°E   40°E

© Oxford University Press
Zenithal Equal Area Projection

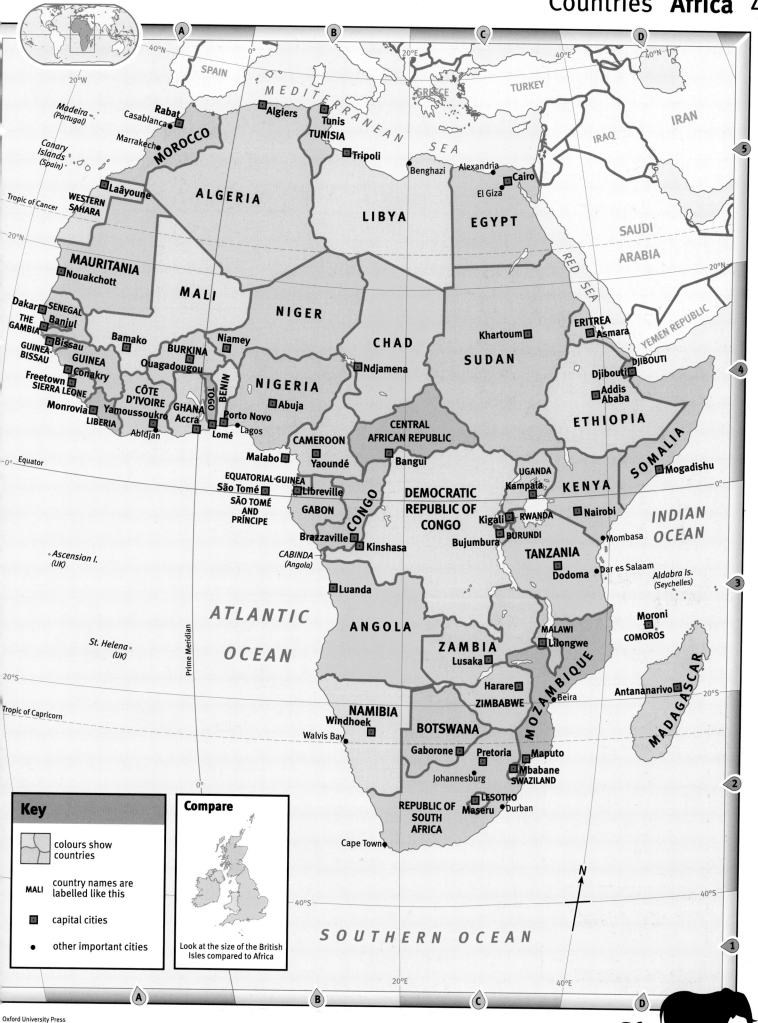

A B C D

40°N

SPAIN

MEDITERRANEAN

GREECE TURKEY

IRAN

IRAQ

20°W

0°

20°E

40°N

5

Madeira
(Portugal)

Rabat
Casablanca
Marrakech
MOROCCO

Algiers

Tunis
TUNISIA

Tripoli

SEA

Benghazi Alexandria
Cairo
El Giza

SAUDI

ARABIA

Canary
Islands
(Spain)

Laâyoune
WESTERN
SAHARA

ALGERIA

LIBYA

EGYPT

Tropic of Cancer

20°N

20°N

MAURITANIA
Nouakchott

MALI

NIGER

CHAD

Khartoum

SUDAN

ERITREA
Asmara

YEMEN REPUBLIC

4

Dakar SENEGAL
THE Banjul
GAMBIA
GUINEA-
BISSAU Bissau Bamako
GUINEA BURKINA
Conakry Ouagadougou
Freetown
SIERRA LEONE

Niamey

Ndjamena

BENIN

NIGERIA

Abuja

DJIBOUTI
Djibouti
Addis
Ababa

RED SEA

Monrovia Yamoussoukro
LIBERIA CÔTE GHANA
D'IVOIRE Accra
Abidjan

TOGO
Porto Novo
Lomé Lagos

ETHIOPIA

Equator

0°

Malabo
Yaoundé

CAMEROON

CENTRAL
AFRICAN REPUBLIC

Bangui

SOMALIA Mogadishu

EQUATORIAL GUINEA
São Tomé
SÃO TOMÉ
AND
PRÍNCIPE

Libreville

GABON

CONGO

DEMOCRATIC
REPUBLIC OF
CONGO

UGANDA
Kampala

KENYA

0°

Kigali RWANDA

Nairobi

INDIAN
OCEAN

Ascension I.
(UK)

Brazzaville
Kinshasa

Bujumbura BURUNDI

TANZANIA

Mombasa

CABINDA
(Angola)

Dodoma Dar es Salaam

Aldabra Is.
(Seychelles)

3

Luanda

St. Helena
(UK)

ATLANTIC

OCEAN

ANGOLA

MALAWI
Lilongwe

Moroni
COMOROS

ZAMBIA

Lusaka

MOZAMBIQUE

MADAGASCAR

20°S

Harare Beira
ZIMBABWE

Antananarivo

20°S

Tropic of Capricorn

Prime Meridian

0°

NAMIBIA
Windhoek

Walvis Bay

BOTSWANA

Gaborone Pretoria
Johannesburg Maseru Durban

Maputo
Mbabane
SWAZILAND

2

Key

colours show
countries

MALI country names are
labelled like this

capital cities

other important cities

Compare

Look at the size of the British
Isles compared to Africa

LESOTHO

REPUBLIC OF
SOUTH
AFRICA

Cape Town

N

40°S

20°S

1

A B C D

20°E

40°E

SOUTHERN OCEAN

Oxford University Press

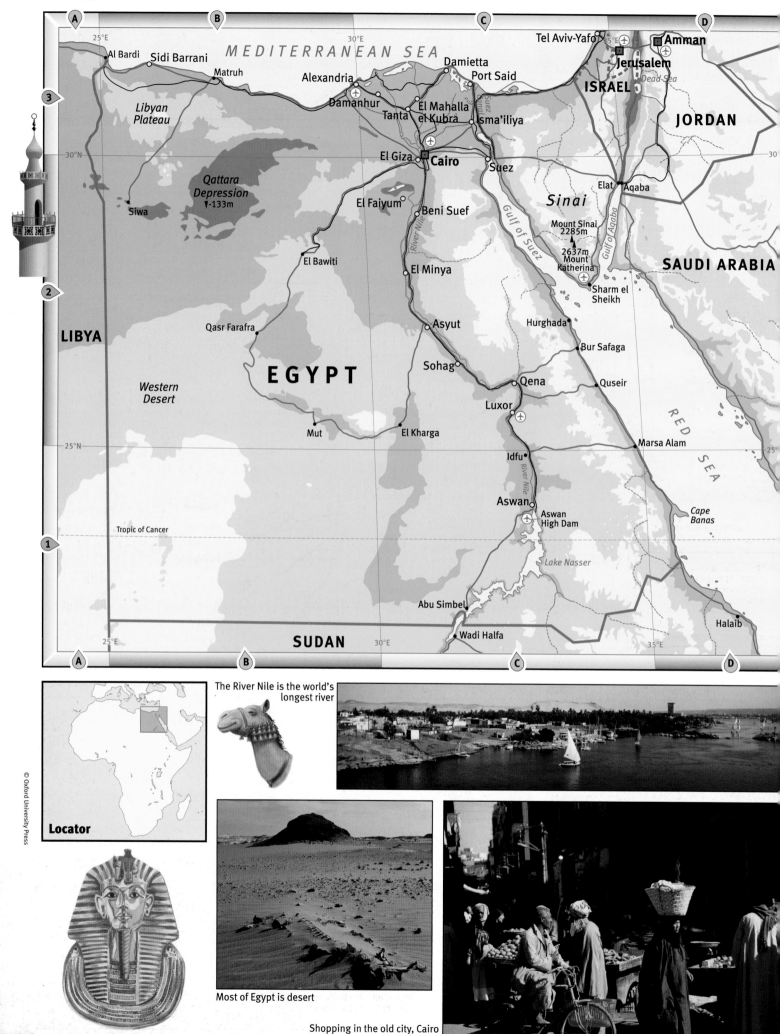

MEDITERRANEAN SEA

Al Bardi
Sidi Barrani
Matruh
Libyan Plateau
25°E
30°E

Damietta
Alexandria
Damanhur
Tanta
El Mahalla el Kubra
Isma'iliya
Port Said

Tel Aviv-Yafo
Amman
Jerusalem
ISRAEL
JORDAN
Dead Sea

El Giza
Cairo
Suez

30°N
30°

Qattara Depression
▼-133m

Siwa

El Faiyum
Beni Suef

Elat
Aqaba
Sinai

Mount Sinai
2285m
▲
2637m
Mount Katherina

SAUDI ARABIA

El Bawiti
El Minya

Sharm el Sheikh

LIBYA

Western Desert

Qasr Farafra

EGYPT

Asyut

Sohag

Hurghada
Bur Safaga

River Nile

Qena
Luxor

Quseir

RED SEA

Mut
El Kharga

Marsa Alam

25°N

Idfu

Tropic of Cancer

Cape Banas

Aswan
Aswan High Dam

River Nile

Lake Nasser

Abu Simbel

Halaib

SUDAN
Wadi Halfa

25°E
30°E
35°E

Locator

The River Nile is the world's longest river

Most of Egypt is desert

Shopping in the old city, Cairo

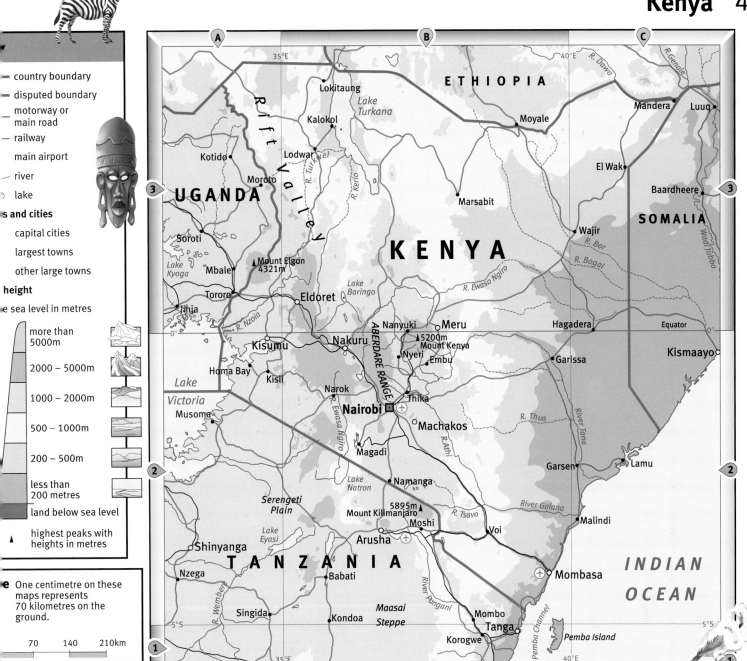

Legend

- country boundary
- disputed boundary
- motorway or main road
- railway
- main airport
- river
- lake

s and cities
- capital cities
- largest towns
- other large towns

height
e sea level in metres

- more than 5000m
- 2000 – 5000m
- 1000 – 2000m
- 500 – 1000m
- 200 – 500m
- less than 200 metres
- land below sea level
- ▲ highest peaks with heights in metres

e One centimetre on these maps represents 70 kilometres on the ground.

| 70 | 140 | 210km |

qual Area Projection

Map labels

UGANDA
Rift Valley
Lokitaung
Lake Turkana
Kalokol
Kotido
Lodwar
R. Turkwel
Moroto
R. Kerio
Soroti
Lake Kyoga
Mount Elgon 4321m
Mbale
Lake Baringo
Tororo
Jinja
Eldoret
R. Nzoia
Kisumu
Nakuru
ABERDARE RANGE
Homa Bay
Kisii
Narok
Ewaso Ngiro
Lake Victoria
Musoma
Nairobi ✈
Magadi
Lake Natron
Namanga
Serengeti Plain
Lake Eyasi
Mount Kilimanjaro 5895m ▲
Moshi
Arusha ✈
Shinyanga
Nzega
TANZANIA
Babati
Singida
Kondoa
Maasai Steppe
River Pangani
Mombo
Tanga
Korogwe
Pemba Channel
Pemba Island

ETHIOPIA
35°E
40°E
R. Dawa
R. Gendle
Moyale
Mandera
Luuq
El Wak
Baardheere
Marsabit
SOMALIA
Wajir
R. Bor
R. Bogal
KENYA
Hagadera
Equator
Wad' Jubba
R. Ewaso Ngiro
Nanyuki
Meru
5200m Mount Kenya ▲
Nyeri
Embu
Garissa
Thika
Machakos
R. Thua
River Tana
Kismaayo
R. Athi
Garsen
Lamu
R. Tsavo
River Galana
Malindi
Voi
Mombasa
INDIAN OCEAN
5°S
40°E

The Maasai people herd cattle in central Kenya

Kilimanjaro is Africa's highest mountain

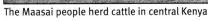

From Mombasa to Nairobi by road takes about 8 hours

Locator

Key

land height in metres above sea level

more than 2000m

1000 – 2000m

500 – 1000m

200 – 500m

less than 200 metres

land below sea level

▲ highest peaks with heights in metres

lake

river

Scale One centimetre on the map represents 400 kilometres on the ground.

0   400   800   1200km

© Oxford University Press
Oblique Mercator Projection

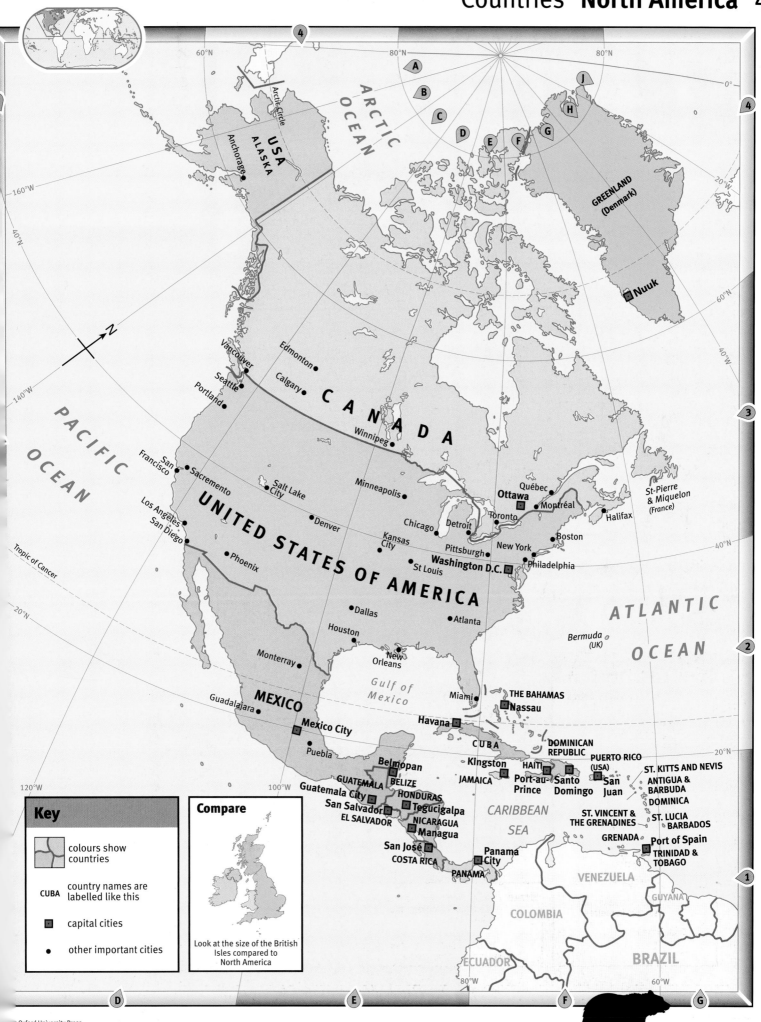

60°N

80°N   80°N

ARCTIC OCEAN

A
B
C
D   E   F   G
H
J

0°

GREENLAND (Denmark)

Arctic Circle

USA ALASKA

Anchorage

160°W

40°N

Nuuk

20°W

60°N

PACIFIC OCEAN

N

Vancouver
Seattle
Portland

Edmonton
Calgary

C A N A D A

Winnipeg

40°W

140°W

3

San Francisco
Sacramento

San
Salt Lake City

Minneapolis

Québec
Ottawa   Montréal

St-Pierre & Miquelon (France)

Los Angeles
San Diego

UNITED STATES OF AMERICA

Denver

Chicago   Detroit   Toronto

Halifax

Phoenix

Kansas City

St Louis

Pittsburgh

New York
Boston

Washington D.C.   Philadelphia

40°N

Tropic of Cancer

20°N

ATLANTIC OCEAN

Dallas

Atlanta

2

Houston

Monterray

New Orleans

Gulf of Mexico

Miami

THE BAHAMAS
Nassau

Bermuda (UK)

120°W

Guadalajara

MEXICO

Mexico City

Puebla

Havana

CUBA

DOMINICAN REPUBLIC

PUERTO RICO (USA)
Kingston   HAITI   San Juan

ST. KITTS AND NEVIS

20°N

100°W

Belmopan

GUATEMALA   BELIZE
Guatemala City   HONDURAS
San Salvador   Tegucigalpa
EL SALVADOR   NICARAGUA
Managua

JAMAICA   Port-au-Prince   Santo Domingo

CARIBBEAN SEA

ANTIGUA & BARBUDA
DOMINICA

ST. VINCENT & THE GRENADINES   ST. LUCIA
BARBADOS

GRENADA   Port of Spain

San José
COSTA RICA

Panama City

TRINIDAD & TOBAGO

PANAMA

VENEZUELA

1

ECUADOR

COLOMBIA

GUYANA

80°W

BRAZIL

60°W

Key

colours show countries

CUBA   country names are labelled like this

capital cities

other important cities

Compare

Look at the size of the British Isles compared to North America

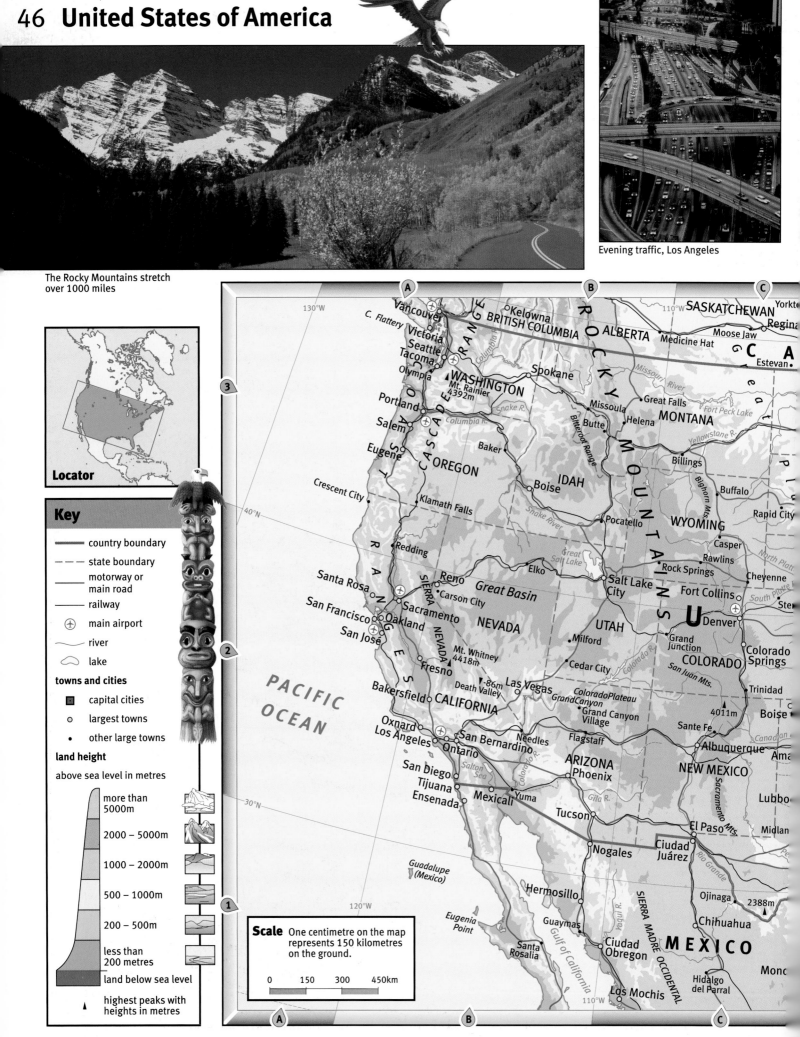

The Rocky Mountains stretch over 1000 miles

Evening traffic, Los Angeles

Locator

Key

——	country boundary
– – –	state boundary
——	motorway or main road
——	railway
✈	main airport
～	river
◠	lake

towns and cities

■	capital cities
○	largest towns
•	other large towns

land height

above sea level in metres

| more than 5000m |
| 2000 – 5000m |
| 1000 – 2000m |
| 500 – 1000m |
| 200 – 500m |
| less than 200 metres |
| land below sea level |
| ▲ highest peaks with heights in metres |

Scale One centimetre on the map represents 150 kilometres on the ground.

0   150   300   450km

© Oxford University Press
Conical Orthomorphic Projection

Wyoming has fewer people than any other state in the USA

Manhattan skyline, New York City

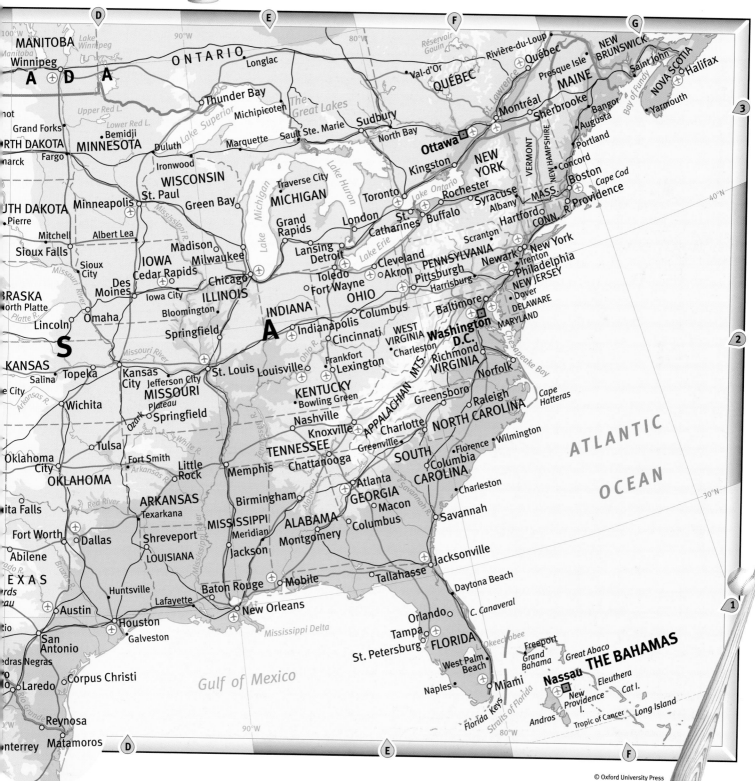

Key

———	country boundary
– – –	disputed boundary
———	motorway or main road
———	railway
⊕	main airport
∿	river
◡	lake

land height

above sea level in metres

more than 5000m

2000 – 5000m

1000 – 2000m

500 – 1000m

200 – 500m

less than 200 metres

land below sea level

▲ highest peaks with heights in metres

towns and cities

◼ capital cities

○ largest towns

• other large towns

Locator

Fishing boats in St. Lucia

Catamarans in the Virgin Islands

FLORIDA
Daytona Beach
Orlando
Cape Canaveral
Tampa
St. Petersburg
L. Okeechobee
West Palm Beach
Miami
Freeport
Grand Bahama
Marsh Harbour
Great Abaco
New Providence Island
Governor's Harbour
Eleuthera
25°N
Florida Keys
Key West
Straits of Florida
Andros Town
Nassau
Andros
Cat Island
San Salvado
THE BAHAMAS
Tropic of Cancer
Archipiélago de Sabana
Great Exuma
Long Islan
Crook I.
Havana
Matanzas
Güines
Sagua la Grande
Acklins Island
Pinar del Río
Santa Clara
Archipiélago de Camagüey
Le Fé
Cienfuegos
Morón
Sancti Spíritus
Nuevitas
Cabo San Antonio
Trinidad
Ciego de Ávila
CUBA
Isla de la Juventud
Camagüey
Victoria de las Tunas
Holguín
G
r
Bayamo
Guantánam
e
Manzanillo
Sierra Maestra
Santiago de Cuba
a
20°N
Windw
Cayman Islands (UK)
George Town
Grand Cayman
t
Jérémi
Montego Bay
e
South Negril Point
JAMAICA
r
Black River
Spanish Town
Kingston
CARIBB
15°N
80°W
75°W

Scale One centimetre on the map represents 80 kilometres on the ground.

0   80   160   240km

© Oxford University Press

Zenithal Equidistant Projection

Main Map (Caribbean)

A T L A N T I C

O C E A N

70°W
65°W

25°N

Tropic of Cancer

Mayaguana

Caicos Passage

Caicos Islands

Grand Turk

Turks and Caicos Is. (UK)

Turks Islands

20°N

Hispaniola

W e s t

Port-de-Paix

Cap Haïtien

Santiago
San Francisco

La Vega

DOMINICAN REPUBLIC

San Juan

Charlotte Amalie

Road Town

Virgin Is. (UK)

The Valley
Anguilla (UK)
Saint Martin (Fr.)

ANTIGUA AND BARBUDA
Barbuda

Codrington

I n d i e s

Leeward

HAITI

Cordillera Central
3175m

San Pedro

La Romana

Aguadilla

Mayagüez

Santo Domingo

Mona Passage

Caguas

Ponce

Puerto Rico (USA)

Virgin Is. (USA)

St. Croix (USA)

St. Maarten (Neths)

St. Kitts

Nevis

Basseterre
ST. KITTS AND NEVIS

Antigua
St. John's

Grande Terre
Guadeloupe (Fr.)

Plymouth
Montserrat (UK)

Pointe-á-Pitre

I s l a n d s

Port-au-Prince

Jacmel

Barahona

Cabo Beata

Basse-Terre

Marie Galente

DOMINICA

Roseau

Lesser Antilles

A n t i l l e s

S E A

N

15°N

Martinique (Fr.)

Fort-de-France

Castries
ST. LUCIA
Vieux Fort

BARBADOS
Bridgetown

ST. VINCENT AND THE GRENADINES

St. Vincent

Kingstown

Windward Islands

L e s s e r   A n t i l l e s

St. George's
GRENADA

Tobago

Punta Gallinas

Oranjestad
Aruba (Neths.)

Curacao (Neths.)

Bonaire (Neths.)

Willemstad

Netherlands Antilles

Punto Fijo

Golfo de Venezuela

Isla Margarita

Port of Spain
TRINIDAD AND TOBAGO

Carúpano

Trinidad
San Fernando

65°W
60°W
70°W

Inset Map (St. Lucia)

St. Lucia

61°W

Saint Lucia Channel

Pte. du Cap
Pigeon Pt.

Gros Islet

Choc Bay

Monchy

Cape Marquis

Grande Rivière

Marquis

Castries

Grande Anse Bay

Babonneau

14°N

La Croix Mairgot

La Sorcière 675m

Marigot

Grande Rivière

Anse la Raye

La Caye

Canaries

Dennery

Praslin Bay

Mt. Gimie 950m

R. Roseau

R. Mabouya

R. Troumassee

Soufrière

Mon Repos

Micoud

Gros Piton 798m

Desruisseaux

Choiseul

Augier

Saltibus Pt.

Laborie

Vieux Fort

Cape Moule à Chique

61°W

Saint Vincent Passage

CARIBBEAN SEA

Cocos Islands

River Magdalena

Lake Maracaibo

River Orinoco

L l a n o s

G U I A N A

Mt. Roraima
▲2810m

H I G H L A N D S

ATLANTIC

OCEAN

Equator

Cotopaxi
5896m ▲

Chimborazo
6310m ▲

Galapagos Islands

A

River Amazon

River Negro

River Amazon

Amazon Basin

River Tapajos

Rocas Island

S e l v a s

River Madeira

N

River Ucayali

River Tocantins

Mato

Grosso

River São Francisco

D

Lake
Titicaca

B R A Z I L I A N

Lake
Poopo

E

HIGHLANDS

PACIFIC

OCEAN

Atacama Desert

River Pilcomayo

River Paraguay

20°S

20°S

Tropic of Capricorn

S

▲6908m
Ojos del
Salado

Gran Chaco

River Paraná

River Uruguay

Aconcagua
6960m ▲

Juan Fernández Islands

P a m p a s

Rio de la Plata

N

R. Colorado

R. Negro

Chiloé Island

Valdés
Peninsula

A

Patagonia

40°S

40°S

Falkland Islands

Tierra
del Fuego

South Georgia

Cape Horn

S O U T H E R N   O C E A N

ATLANTIC

OCEAN

Key

land height in metres above sea level

	more than 5000m
	2000 – 5000m
	1000 – 2000m
	500 – 1000m
	200 – 500m
	less than 200 metres
	land below sea level

▲ highest peaks with heights in metres

lake

river

Scale One centimetre on the map represents 350 kilometres on the ground.

0   350   700   1050km

N

CARIBBEAN SEA

ATLANTIC OCEAN

NICARAGUA

COSTA RICA

PANAMA

Barranquilla
Maracaibo
Caracas
Valencia
VENEZUELA
Medellin
Cali □**Bogota**
COLOMBIA

Georgetown
GUYANA
Paramaribo
SURINAME
□**Cayenne**
French Guiana
(France)

Equator

Galapagos Islands
(Ecuador)

Quito
ECUADOR
Guayaquil
Iquitos

Belem
Manaus
Fortaleza

Rocas Island
(Brazil)

Trujillo
PERU

B R A Z I L

Recife

□**Lima**

BOLIVIA
La Paz
Santa Cruz
Sucre

Arequipa

Brásília□

Salvador

Belo Horizonte

Rio de Janeiro

20°S

PARAGUAY
Antofagasta
Asunción□
São Paulo
Curitiba

Tropic of Capricorn

PACIFIC OCEAN

Cordoba
Rosario

Porto Alegre

URUGUAY
Santiago□
Buenos Aires□ □**Montevideo**
ARGENTINA
Concepcion
Mar del Plata

Juan Fernandez Is.
(Chile)

ATLANTIC OCEAN

N

100°W

40°S

Stanley□
Falkland Islands
(UK)

South Georgia
(UK)

Punta Arenas

SOUTHERN OCEAN

20°W

40°W

60°S

Key

colours show countries

PERU country names are labelled like this

□ capital cities

• other important cities

Compare

Look at the size of the British Isles compared to South America

Brazil, Peru, Bolivia

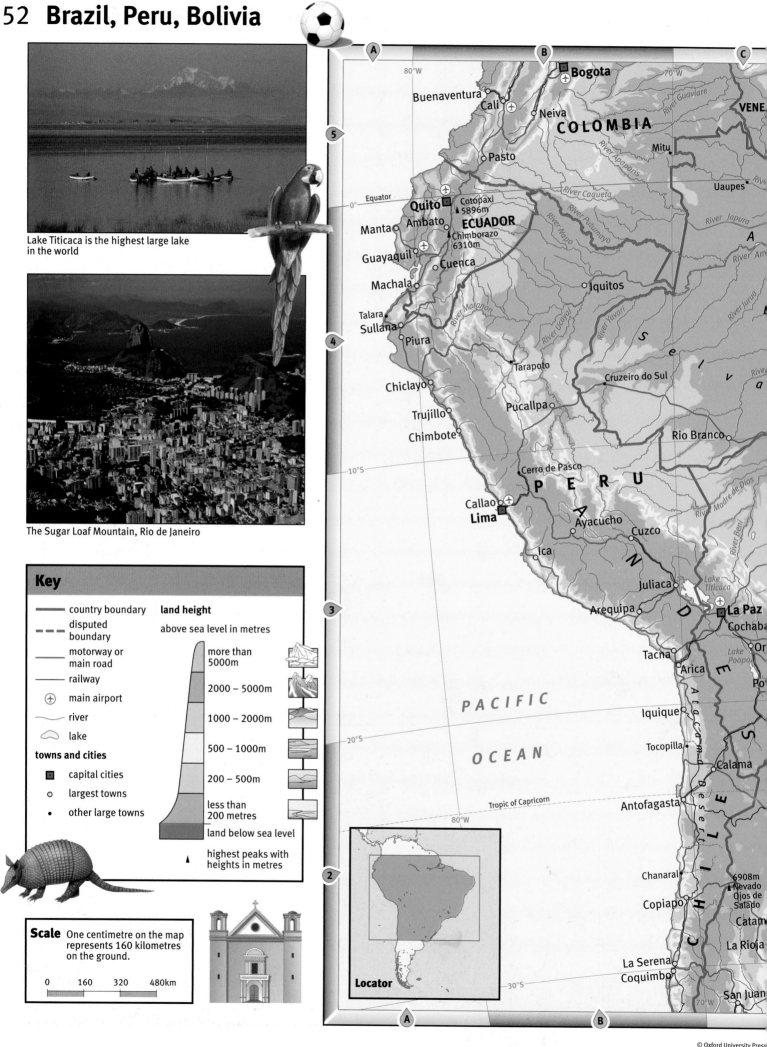

Lake Titicaca is the highest large lake in the world

The Sugar Loaf Mountain, Rio de Janeiro

Key

- country boundary
- disputed boundary
- motorway or main road
- railway
- main airport
- river
- lake

towns and cities

- capital cities
- largest towns
- other large towns

land height

above sea level in metres

- more than 5000m
- 2000 – 5000m
- 1000 – 2000m
- 500 – 1000m
- 200 – 500m
- less than 200 metres
- land below sea level
- highest peaks with heights in metres

Scale One centimetre on the map represents 160 kilometres on the ground.

0 160 320 480km

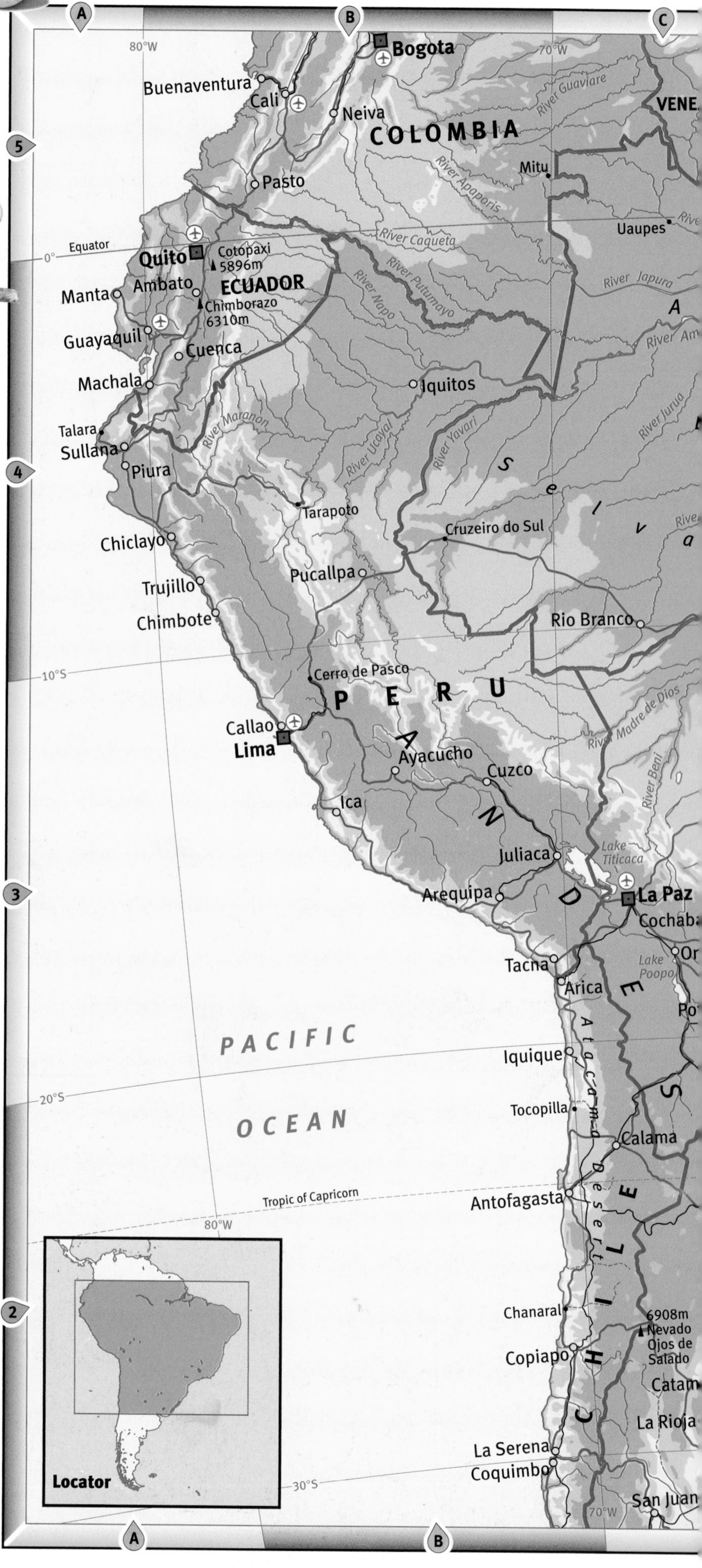

A
B
C
80°W
70°W
Bogota
Buenaventura
Cali
Neiva
COLOMBIA
River Guaviare
VENE
River Apaporis
Mitu
Pasto
Uaupes
River Caqueta
River Putumayo
River Napo
Equator
0°
Quito
Cotopaxi 5896m
ECUADOR
Ambato
Manta
Chimborazo 6310m
River Japura
Guayaquil
Cuenca
River Am
Machala
Iquitos
Talara
Sullana
River Maranon
River Ucayali
River Yavari
River Jurua
Piura
S e l v a
Tarapoto
Cruzeiro do Sul
Chiclayo
Trujillo
Pucallpa
Chimbote
Rio Branco
10°S
Cerro de Pasco
P E R U
River Madre de Dios
Callao
Lima
A N D E S
Ayacucho
Cuzco
River Beni
Ica
Juliaca
Lake Titicaca
Arequipa
La Paz
Cochaba
Lake Poopo
Tacna
Arica
PACIFIC
OCEAN
Atacama Desert
Iquique
20°S
Tocopilla
Calama
Tropic of Capricorn
Antofagasta
80°W
C H I L E
Chanaral
6908m Nevado Ojos de Salado
Copiapo
Catam
La Rioja
La Serena
Coquimbo
30°S
70°W
San Juan
Locator
5
4
3
2

© Oxford University Press
Transverse Mercator Projection

ATLANTIC OCEAN

ATLANTIC OCEAN

Equator 0°

Tropic of Capricorn

60°W
50°W
40°W
30°S

GUYANA

SURINAME

French Guiana (France)

Boa Vista

Serra Tumucumaque

Macapa

Ilha de Marajo

Bragança

Belem

São Luis

Parnaiba

Sobral

Fortaleza

Mossoro

Natal

Joao Pessoa

Campina Grande

Recife

Caruaru

Maceio

Aracaju

Alagoinhas

Salvador

Jequie

Ilheus

Vitoria da Conquista

Teofilo Otoni

Governador Valadares

Linhares

Vitoria

Campos

Rio de Janeiro

Nova Iguaçu

Santo Andre

Santos

Barbacena

Juiz de Fora

Belo Horizonte

Caratinga

Mount Itambe 2033m

Montes Claros

Petrolina

Feira de Santana

Barreiras

Brasilia

Anapolis

Goiania

Rio Verde

Uberlandia

Uberaba

Ribeirao Preto

São Jose do Rio Preto

Araraquara

Bauru

Campinas

São Paulo

Ponta Grossa

Curitiba

Paranagua

Itajai

Florianopolis

Caxias do Sul

Passo Fundo

Santa Maria

Uruguaiana

Porto Alegre

Pelotas

Rio Grande

URUGUAY

Maringa

Foz do Iguaçu

Asuncion

PARAGUAY

Formosa

Resistencia

Corrientes

Posadas

Pedro Juan Caballero

Dourados

Campo Grande

Sa. de Maracaju

Corumba

Santa Cruz

BOLIVIA

Salvador

Miguel de Tucuman

Santiago del Estero

ARGENTINA

Santa Fe

Parana

Concordia

Cordoba

Cuiaba

Caceres

Rondonopolis

Mato Grosso

BRAZIL

BRAZILIAN HIGHLANDS

Chapada Diamantina

Juazeiro do Norte

Barra do Corda

Imperatriz

Maraba

Araguaina

Altamira

Tucurui

Codo

Teresina

Caxias

Bacabal

Cameta

Santarem

Itaituba

Manaus

Manacapuru

Coari

Porto Velho

Ariquemes

Balbina Reservoir

Mouths of the Amazon

River Amazon

R. Xingu

River Iriri

River Araguaia

River Tocantins

River Parnaiba

River São Francisco

River Jequitinhonha

River Paranaíba

River Teles Pires

River Tapajos

River Madeira

River Aripuana

River Juruena

River Guapore

River Branco

River Juruena

River Paraguay

River Parana

River Pilcomayo

River Bermejo

River Salado

River Uruguay

Gran Chaco

Lagoa dos Patos

10°S

20°S

© Oxford University Press

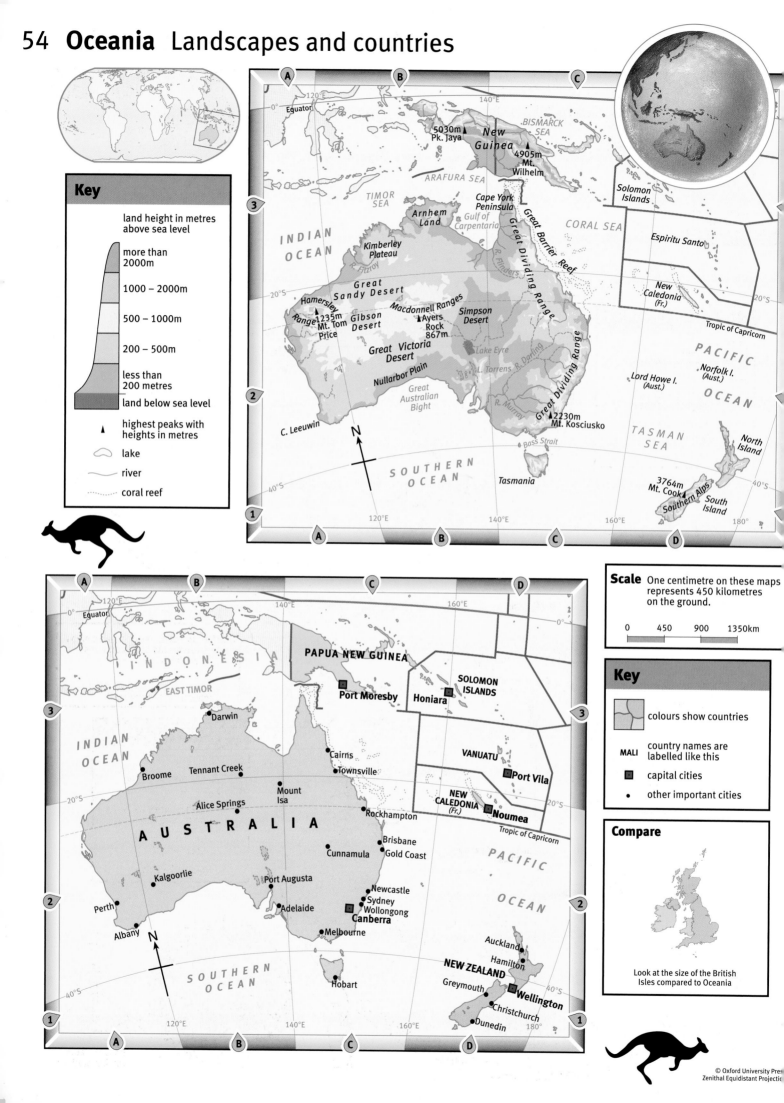

Key

land height in metres above sea level

- more than 2000m
- 1000 – 2000m
- 500 – 1000m
- 200 – 500m
- less than 200 metres
- land below sea level
- ▲ highest peaks with heights in metres
- lake
- river
- coral reef

Map 1 (Landscapes):

A B C

120°E 140°E

0° Equator

Pk. Jaya 5030m ▲ **New Guinea** 4905m ▲ Mt. Wilhelm

BISMARCK SEA

ARAFURA SEA

TIMOR SEA

Solomon Islands

CORAL SEA

Cape York Peninsula

Arnhem Land

Gulf of Carpentaria

Great Dividing Range

Great Barrier Reef

INDIAN OCEAN

Kimberley Plateau

R. Fitzroy

Espiritu Santo

New Caledonia (Fr.)

Great Sandy Desert

Hamersley Range ▲1235m Mt. Tom Price

Gibson Desert

Macdonnell Ranges

▲ Ayers Rock 867m

Simpson Desert

Lake Eyre

R. Flinders

Tropic of Capricorn

PACIFIC

Great Victoria Desert

L. Torrens R. Darling

Great Dividing Range

Lord Howe I. (Aust.)

Norfolk I. (Aust.)

OCEAN

Nullarbor Plain

Great Australian Bight

R. Murray

▲2230m Mt. Kosciusko

TASMAN SEA

North Island

C. Leeuwin

N

Bass Strait

Tasmania

SOUTHERN OCEAN

3764m Mt. Cook ▲ Southern Alps

South Island

120°E 140°E 160°E 180°

Scale

One centimetre on these maps represents 450 kilometres on the ground.

0 450 900 1350km

Key

- colours show countries
- **MALI** country names are labelled like this
- capital cities
- other important cities

Map 2 (Countries):

A B C D

120°E 140°E 160°E

0° Equator

INDONESIA

EAST TIMOR

PAPUA NEW GUINEA

□ Port Moresby

SOLOMON ISLANDS

□ Honiara

INDIAN OCEAN

Darwin

Broome Tennant Creek

Cairns

Townsville

Mount Isa

Alice Springs

AUSTRALIA

Rockhampton

VANUATU

□ Port Vila

NEW CALEDONIA (Fr.)

□ Noumea

Tropic of Capricorn

Brisbane

Cunnamula Gold Coast

Kalgoorlie

Port Augusta

Newcastle

Sydney

Wollongong

Canberra

Adelaide

PACIFIC OCEAN

Perth

Melbourne

Albany

N

Hobart

SOUTHERN OCEAN

Auckland

Hamilton

NEW ZEALAND

Greymouth

□ **Wellington**

Christchurch

Dunedin

120°E 140°E 160°E 180°

Compare

Look at the size of the British Isles compared to Oceania

© Oxford University Press
Zenithal Equidistant Projection

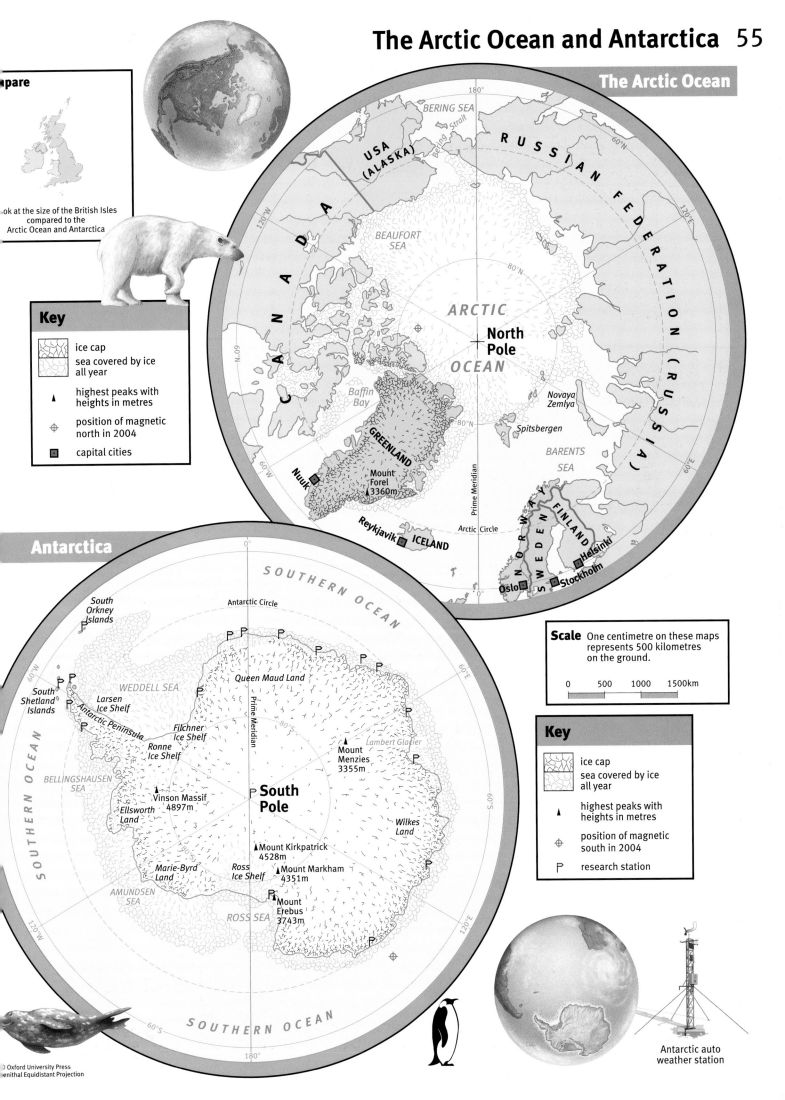

The Arctic Ocean

BERING SEA
Bering Strait
180°
USA (ALASKA)
RUSSIAN FEDERATION (RUSSIA)
60°N
120°W
120°E
CANADA
BEAUFORT SEA
80°N
60°N
60°N
ARCTIC
North Pole
OCEAN
Baffin Bay
80°N
Novaya Zemlya
Spitsbergen
GREENLAND
BARENTS SEA
Mount Forel ▲3360m
Nuuk
Prime Meridian
Arctic Circle
NORWAY
SWEDEN
FINLAND
Reykjavik ICELAND
Helsinki
Oslo
Stockholm
60°E

Key

- ice cap
- sea covered by ice all year
- ▲ highest peaks with heights in metres
- ⊕ position of magnetic north in 2004
- ▣ capital cities

pare

ok at the size of the British Isles compared to the Arctic Ocean and Antarctica

Antarctica

South Orkney Islands
Antarctic Circle
SOUTHERN OCEAN
0°
60°W
60°E
South Shetland Islands
WEDDELL SEA
Queen Maud Land
P P P P P P
P
Larsen Ice Shelf
Antarctic Peninsula
SOUTHERN OCEAN
Filchner Ice Shelf
Prime Meridian
80°S
Lambert Glacier
BELLINGSHAUSEN SEA
Ronne Ice Shelf
Mount Menzies 3355m
P
Vinson Massif ▲4897m
South Pole
Ellsworth Land
Wilkes Land
60°S
Marie-Byrd Land
▲Mount Kirkpatrick 4528m
Ross Ice Shelf
▲Mount Markham 4351m
P
AMUNDSEN SEA
P ▲Mount Erebus 3743m
120°W
ROSS SEA
120°E
P
60°S
SOUTHERN OCEAN
180°

Scale
One centimetre on these maps represents 500 kilometres on the ground.

0 500 1000 1500km

Key

- ice cap
- sea covered by ice all year
- ▲ highest peaks with heights in metres
- ⊕ position of magnetic south in 2004
- P research station

Antarctic auto weather station

© Oxford University Press
enithal Equidistant Projection

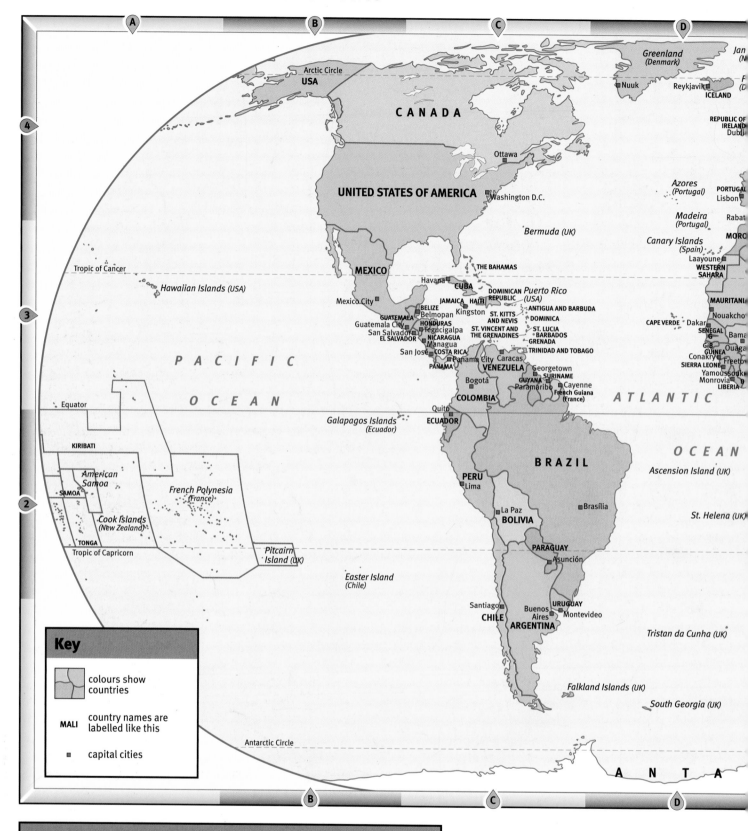

A

B

C

D

Greenland
(Denmark)

Jan
(N)

Arctic Circle

USA

Nuuk

Reykjavik

F
(D)

ICELAND

C A N A D A

REPUBLIC OF
IRELAND
Dubli

4

Ottawa

UNITED STATES OF AMERICA

Washington D.C.

Azores
(Portugal)

PORTUGAL
Lisbon

Bermuda (UK)

Madeira
(Portugal)

Rabat

MORO

Tropic of Cancer

Canary Islands
(Spain)

Hawaiian Islands (USA)

MEXICO

THE BAHAMAS

Havana

Laayoune

WESTERN
SAHARA

Mexico City

CUBA

DOMINICAN *Puerto Rico*
REPUBLIC *(USA)*

JAMAICA **HAITI**

MAURITANIA

CAPE VERDE Dakar

3

GUATEMALA **BELIZE**
Belmopan
Guatemala City **HONDURAS**
San Salvador **NICARAGUA**
EL SALVADOR Managua
San José **COSTA RICA**

Kingston

ST. KITTS
AND NEVIS

ANTIGUA AND BARBUDA

DOMINICA

ST. LUCIA
BARBADOS
GRENADA

Nouakcho

SENEGAL
G

Bama

G-B

GUINEA Ouaga

Conakry **Freeto**

SIERRA LEONE

Tegucigalpa

Yamoussouk

Monrovia **LIBERIA**

PANAMA

TRINIDAD AND TOBAGO

Panama City Caracas

VENEZUELA Georgetown

Bogotá **GUYANA** **SURINAME**

Paramaribo

Cayenne

COLOMBIA *French Guiana*
(France)

A T L A N T I C

Quito

ECUADOR

Galapagos Islands
(Ecuador)

O C E A N

P A C I F I C

Equator

Ascension Island (UK)

KIRIBATI

B R A Z I L

American
Samoa

French Polynesia
(France)

PERU

Lima

O C E A N

2

SAMOA

La Paz

Brasília

St. Helena (UK)

BOLIVIA

Cook Islands
(New Zealand)

PARAGUAY

TONGA

Tropic of Capricorn

Pitcairn
Island (UK)

Asunción

Easter Island
(Chile)

URUGUAY

Santiago

Buenos
Aires Montevideo

Tristan da Cunha (UK)

CHILE

ARGENTINA

Key

colours show
countries

MALI country names are
labelled like this

Falkland Islands (UK)

South Georgia (UK)

■ capital cities

Antarctic Circle

A N T A

B

C

D

Eckert IV Projection

Abbreviations

A	ALBANIA	CZ	CZECH REPUBLIC	Q	QATAR
AR	ARMENIA	G	THE GAMBIA	R	ROMANIA
AU	AUSTRIA	G-B	GUINEA-BISSAU	S	SLOVAKIA
AZ	AZERBAIJAN	H	HUNGARY	SL	SLOVENIA
B	BELGIUM	IS	ISRAEL	SM	SERBIA AND MONTENEGRO
BE	BENIN	L	LEBANON	SW	SWITZERLAND
BH	BOSNIA-HERZEGOVINA	LI	LITHUANIA	T	TAJIKISTAN
BR	BRUNEI	LU	LUXEMBOURG	TU	TURKMENISTAN
BU	BURKINA	M	FORMER YUGOSLAV	U	UGANDA
C	CROATIA		REPUBLIC OF MACEDONIA	UAE	UNITED ARAB EMIRATES
CAR	CENTRAL AFRICAN REPUBLIC	N	NETHERLANDS	ZIM	ZIMBABWE

North America

South America

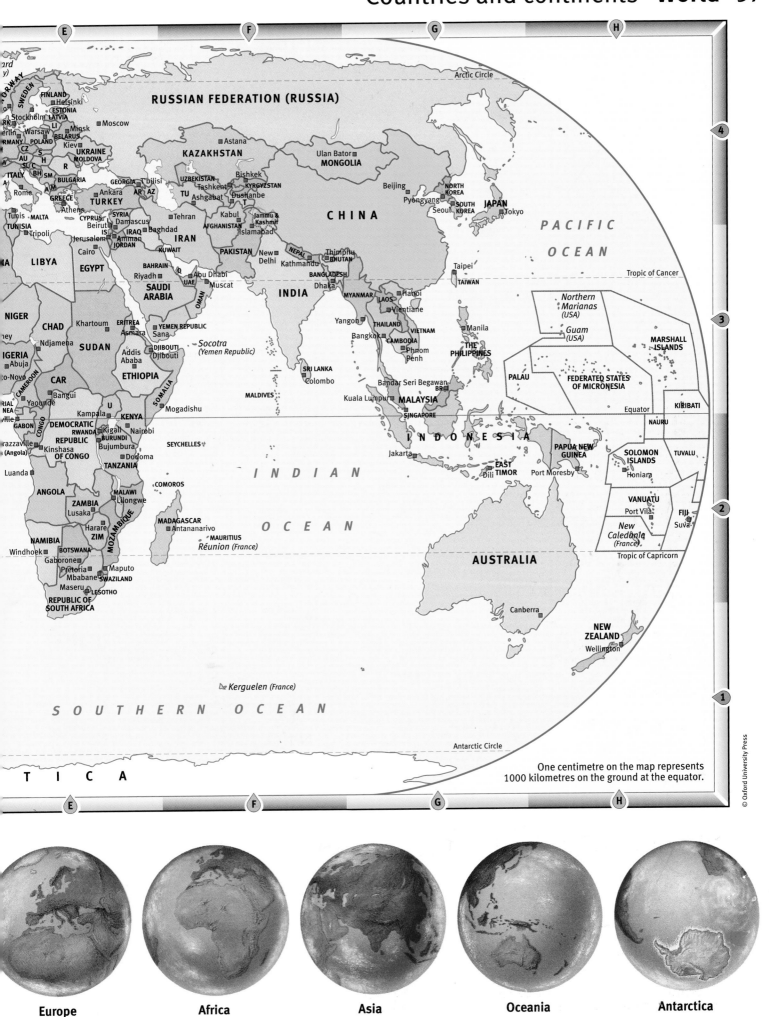

One centimetre on the map represents
1000 kilometres on the ground at the equator.

Europe   **Africa**   **Asia**   **Oceania**   **Antarctica**

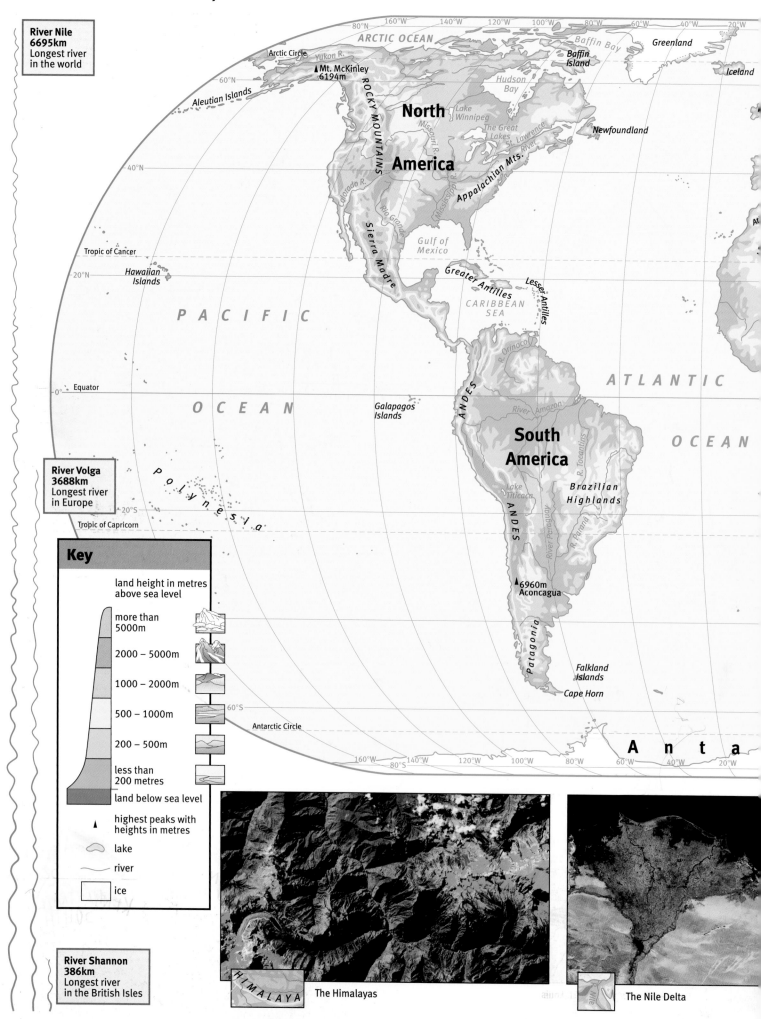

**River Nile
6695km**
Longest river
in the world

ARCTIC OCEAN

Arctic Circle

Yukon R.

▲ Mt. McKinley
6194m

Baffin Bay

Greenland

Baffin
Island

Iceland

Aleutian Islands

ROCKY MOUNTAINS

60°N

Hudson
Bay

North

Lake
Winnipeg

The Great
Lakes

St. Lawrence

Newfoundland

America

Missouri R.

Mississippi R.

Appalachian Mts.

St. Lawrence River.

40°N

Colorado R.

Rio Grande

Sierra Madre

Gulf of
Mexico

Tropic of Cancer

20°N

Hawaiian
Islands

Greater Antilles

Lesser Antilles

CARIBBEAN
SEA

P A C I F I C

R. Orinoco

A T L A N T I C

Equator

0°

O C E A N

Galapagos
Islands

ANDES

River Amazon

O C E A N

South

America

R. Tocantins

**River Volga
3688km**
Longest river
in Europe

P o l y n e s i a

Lake
Titicaca

**Brazilian
Highlands**

ANDES

River Paraguay

R. Paraná

20°S

Tropic of Capricorn

Key

land height in metres
above sea level

more than
5000m

2000 – 5000m

1000 – 2000m

500 – 1000m

200 – 500m

less than
200 metres

land below sea level

ANDES

▲6960m
Aconcagua

Patagonia

Falkland
Islands

Cape Horn

60°S

Antarctic Circle

A n t a

160°W 140°W 120°W 100°W 80°W 60°W 40°W 20°W

80°S

▲ highest peaks with
heights in metres

lake

river

ice

HIMALAYA

The Himalayas

Nile

The Nile Delta

**River Shannon
386km**
Longest river
in the British Isles

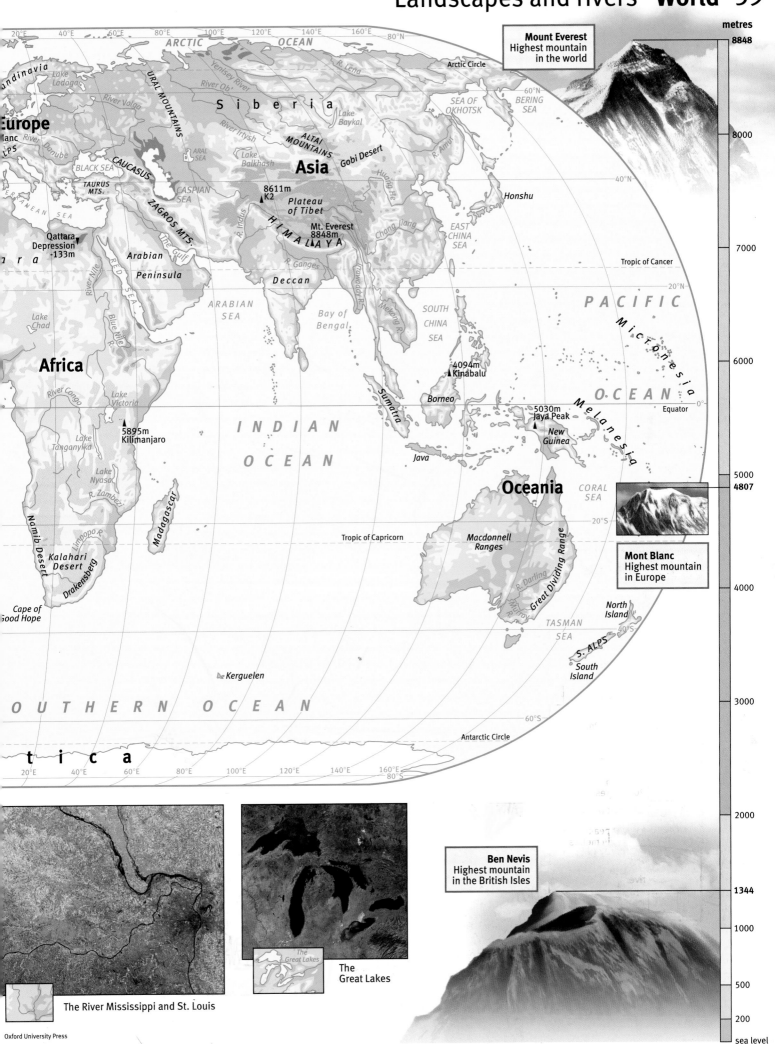

metres

Mount Everest
Highest mountain in the world

8848

8000

7000

Tropic of Cancer

6000

5000

4807

Mont Blanc
Highest mountain in Europe

4000

3000

2000

Ben Nevis
Highest mountain in the British Isles

1344

1000

500

200

sea level

20°E 40°E 60°E 80°E 100°E 120°E 140°E 160°E 80°N

ARCTIC   OCEAN
Arctic Circle

Scandinavia
Lake Ladoga
River Ob'
Yenisey River
R. Lena
60°N
SEA OF OKHOTSK
BERING SEA

Europe

URAL MOUNTAINS

River Volga

S i b e r i a

Lake Baykal

ALTAI MOUNTAINS

River Irtysh

Lake Balkhash

Asia

Gobi Desert

40°N

Honshu

Mont Blanc
ALPS
River Danube

CAUCASUS
BLACK SEA
TAURUS MTS.
CASPIAN SEA

ARAL SEA

8611m
K2

Plateau of Tibet

Huang He

ZAGROS MTS.

Qattara Depression -133m

The Gulf

R. Indus

H I M A L A Y A

Mt. Everest 8848m

Chang Jiang

EAST CHINA SEA

Tropic of Cancer

20°N

Arabian Peninsula

RED SEA

Blue Nile R.

River Nile

Deccan

R. Ganges

Irrawaddy R.

Mekong R.

SOUTH CHINA SEA

P A C I F I C

M i c r o n e s i a

ARABIAN SEA

Bay of Bengal

Lake Chad

Africa

River Congo

Lake Victoria

Sumatra

4094m
Kinabalu

Borneo

O C E A N

Equator 0°

5030m
Jaya Peak

M e l a n e s i a

5895m
Kilimanjaro

Lake Tanganyika

I N D I A N

Java

New Guinea

Lake Nyasa

R. Zambezi

O C E A N

CORAL SEA

Madagascar

Oceania

20°S

Namib Desert

Limpopo R.

Kalahari Desert

Drakensberg

Tropic of Capricorn

Macdonnell Ranges

Great Dividing Range

North Island

Cape of Good Hope

R. Darling

Murray

TASMAN SEA

40°S

S. ALPS
South Island

Kerguelen

S O U T H E R N   O C E A N

60°S

Antarctic Circle

t i c a

20°E 40°E 60°E 80°E 100°E 120°E 140°E 160°E 80°S

The River Mississippi and St. Louis

The Great Lakes

60 **World** Climates

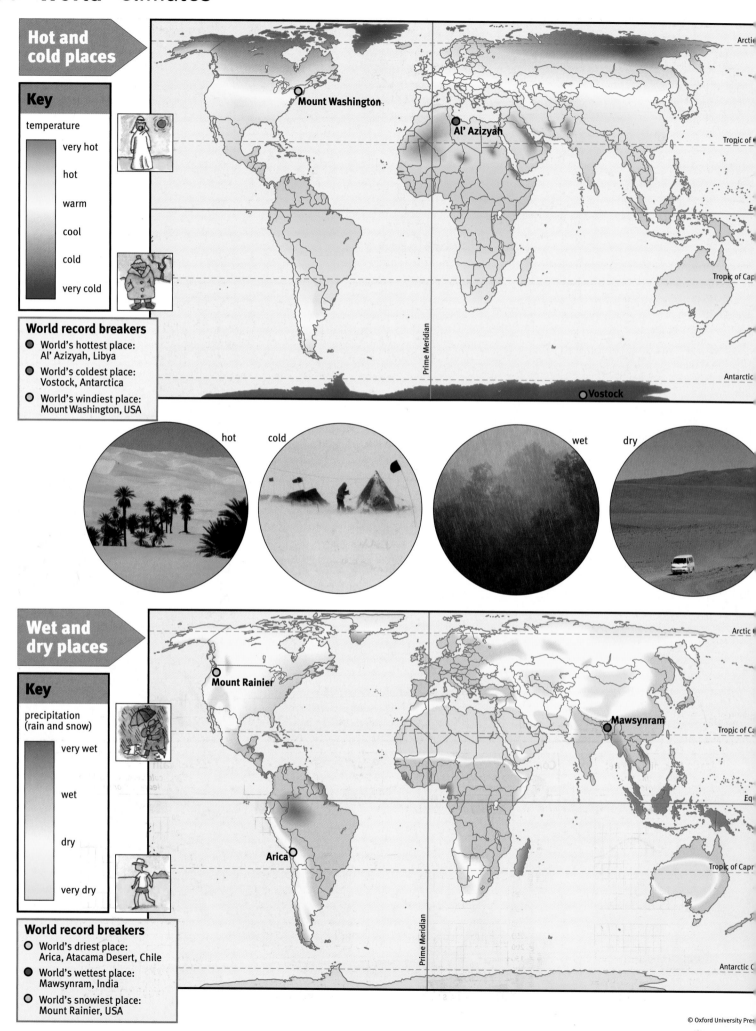

Hot and cold places

Key

temperature

- very hot
- hot
- warm
- cool
- cold
- very cold

World record breakers

- ◉ World's hottest place:
  Al' Azizyah, Libya
- ◉ World's coldest place:
  Vostock, Antarctica
- ○ World's windiest place:
  Mount Washington, USA

Mount Washington

Al' Azizyah

Vostock

Arctic

Tropic of

Equator

Tropic of Cap

Antarctic

Prime Meridian

hot · cold · wet · dry

Wet and dry places

Key

precipitation
(rain and snow)

- very wet
- wet
- dry
- very dry

World record breakers

- ○ World's driest place:
  Arica, Atacama Desert, Chile
- ◉ World's wettest place:
  Mawsynram, India
- ○ World's snowiest place:
  Mount Rainier, USA

Mount Rainier

Mawsynram

Arica

Arctic

Tropic of Ca

Equator

Tropic of Capr

Antarctic

Prime Meridian

© Oxford University Pres

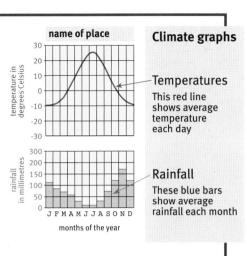

Climate graphs

name of place

temperature in degrees Celsius

Temperatures
This red line shows average temperature each day

rainfall in millimetres

Rainfall
These blue bars show average rainfall each month

months of the year

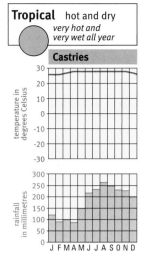

Tropical   hot and dry
very hot and very wet all year

Castries

temperature in degrees Celsius

rainfall in millimetres

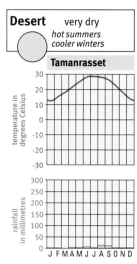

Desert   very dry
hot summers cooler winters

Tamanrasset

temperature in degrees Celsius

rainfall in millimetres

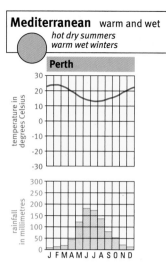

Mediterranean   warm and wet
hot dry summers warm wet winters

Perth

temperature in degrees Celsius

rainfall in millimetres

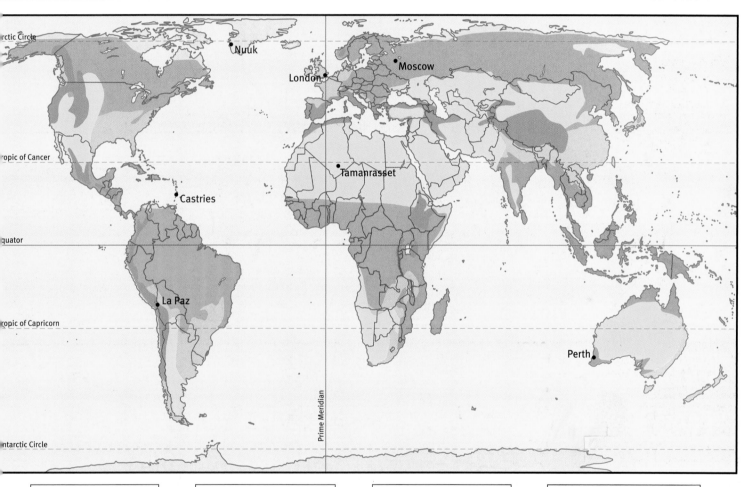

Nuuk

Moscow

London

Tamanrasset

Castries

La Paz

Perth

Arctic Circle

Tropic of Cancer

Equator

Tropic of Capricorn

Antarctic Circle

Prime Meridian

Oxford University Press
Eckert IV projection

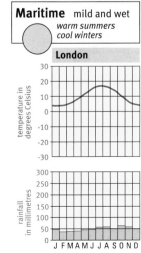

Maritime   mild and wet
warm summers cool winters

London

temperature in degrees Celsius

rainfall in millimetres

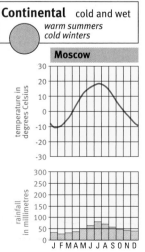

Continental   cold and wet
warm summers cold winters

Moscow

temperature in degrees Celsius

rainfall in millimetres

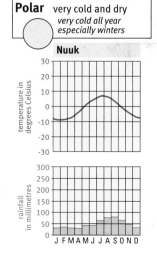

Polar   very cold and dry
very cold all year especially winters

Nuuk

temperature in degrees Celsius

rainfall in millimetres

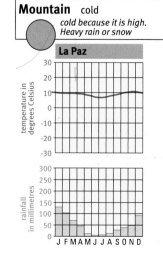

Mountain   cold
cold because it is high. Heavy rain or snow

La Paz

temperature in degrees Celsius

rainfall in millimetres

There are about
6 400 000 000
people in the world.

Population density

ARCTIC OCEAN
Arctic Circle

Chicago
New York
Washington D.C.
Philadelphia
San Francisco
Los Angeles

Tropic of Cancer

Mexico City

PACIFIC
OCEAN

Bogotá

ATLANTIC

OCEAN

Equator

Lima-Callao

Tropic of Capricorn

Rio de Janeiro
São Paulo

Buenos Aires

Antarctic Circle

Key

Population density

people per square kilometre

	over 100
	5–100
	under 5
■	cities with more than six million (6 000 000) people
—	country boundary

Population pyramid

If there were just 100 people in the world,
this is how old they would be:

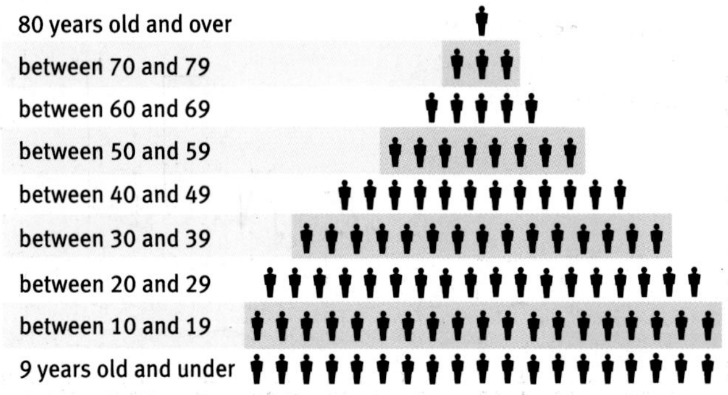

80 years old and over
between 70 and 79
between 60 and 69
between 50 and 59
between 40 and 49
between 30 and 39
between 20 and 29
between 10 and 19
9 years old and under

Where people live

If there were just
100 people in
the world, this
is where they
would live:

- Europe
- Asia
- Africa
- North America
- South America
- Oceania

© Oxford University Press
Eckert IV Projection

millions

Moscow

Paris

Istanbul

Tehran

Cairo

Lahore

Karachi

Delhi

Beijing

Seoul

Nagoya

Tokyo

Osaka

Shanghai

Chongqing

Taipei

Kolkata

Dhaka

Mumbai

Hyderabad

Hong Kong

Bangalore

Chennai

Bangkok

Manila

Lagos

Kinshasa

Jakarta

Johannesburg

ARCTIC OCEAN

Arctic Circle

PACIFIC OCEAN

Tropic of Cancer

INDIAN OCEAN

Equator

Tropic of Capricorn

SOUTHERN OCEAN

Antarctic Circle

20°E 40°E 60°E 80°E 100°E 120°E 140°E 160°E 80°N

60°N

40°N

20°N

0°

20°S

40°S

60°S

- 6250
- 6000
- 5750
- 5500
- 5250
- 5000
- 4750
- 4500
- 4250
- 4000
- 3750
- 3500
- 3250
- 3000
- 2750
- 2500
- 2250
- 2000
- 1750
- 1500
- 1250
- 1000
- 750
- 500
- 250
- 0

Births and deaths

In **2003**... 128 758 963 people were born

☺ ☺ ☺ ☺ ☺ ☺ ☺ ☺ ☺ ☺ ☺ ☺ ☺

and... 55 508 568 people died

☹ ☹ ☹ ☹ ☹

each ☺ represents
10 000 000 births
and each ☹
represents
10 000 000 deaths.

so... **73 250 395** **people were added to the world's population**

Population growth

In the last 50 years, world population has grown very fast.

1200 1300 1400 1500 1600 1700 1800 1900 2000

Oxford University Press

tropical forest

deciduous forest

coniferous forest

ARCTIC OCEAN

Arctic Circle

ARCTIC OCEAN

80°N

160°W   140°W   120°W   100°W   80°W   60°W   40°W   20°W

60°N

40°N

Tropic of Cancer

20°N

PACIFIC

Equator   0°

OCEAN

20°S

Tropic of Capricorn

ATLANTIC

OCEAN

40°S

60°S

Antarctic Circle

80°S   160°W   140°W   120°W   100°W   80°W   60°W   40°W   20°W

Key

	coniferous forest trees have leaves all year
	deciduous forest trees drop their leaves in winter
	tropical forest tall trees growing close together
	savannah tall trees and scattered trees
	temperate grassland prairies, steppes, pampas and veld
	semi desert short grass and small dry bushes
	desert sand and stones with few plants
	tundra moss and bog with some short trees
	ice no plants
	mountains thin soils and steep slopes

desert

semi desert

savannah

temperate grassland

ARCTIC OCEAN

Arctic Circle

60°N

40°N

PACIFIC

Tropic of Cancer

20°N

OCEAN

Equator

INDIAN

OCEAN

20°S

Tropic of Capricorn

40°S

© Oxford University Press

SOUTHERN OCEAN

60°S

Antarctic Circle

20°E 40°E 60°E 80°E 100°E 120°E 140°E 160°E 80°S

tundra

mountains

ice

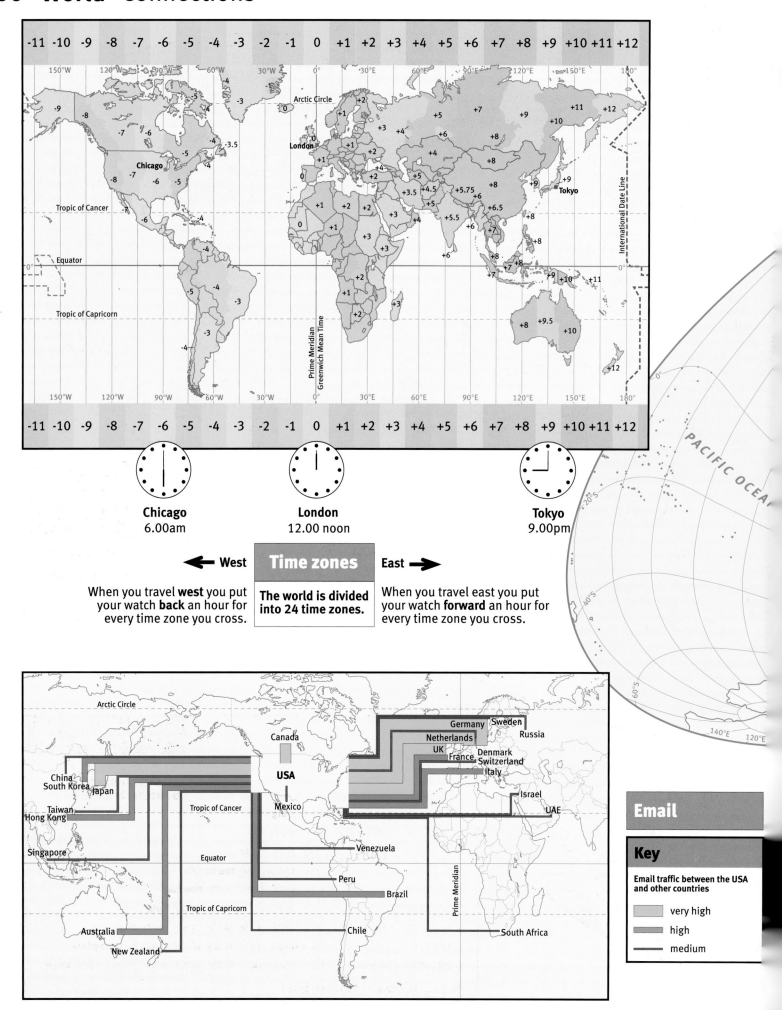

| -11 | -10 | -9 | -8 | -7 | -6 | -5 | -4 | -3 | -2 | -1 | 0 | +1 | +2 | +3 | +4 | +5 | +6 | +7 | +8 | +9 | +10 | +11 | +12 |

Chicago
6.00am

London
12.00 noon

Tokyo
9.00pm

← **West**   **Time zones**   **East** →

When you travel **west** you put your watch **back** an hour for every time zone you cross.

The world is divided into 24 time zones.

When you travel east you put your watch **forward** an hour for every time zone you cross.

Email

Key

Email traffic between the USA and other countries

very high

high

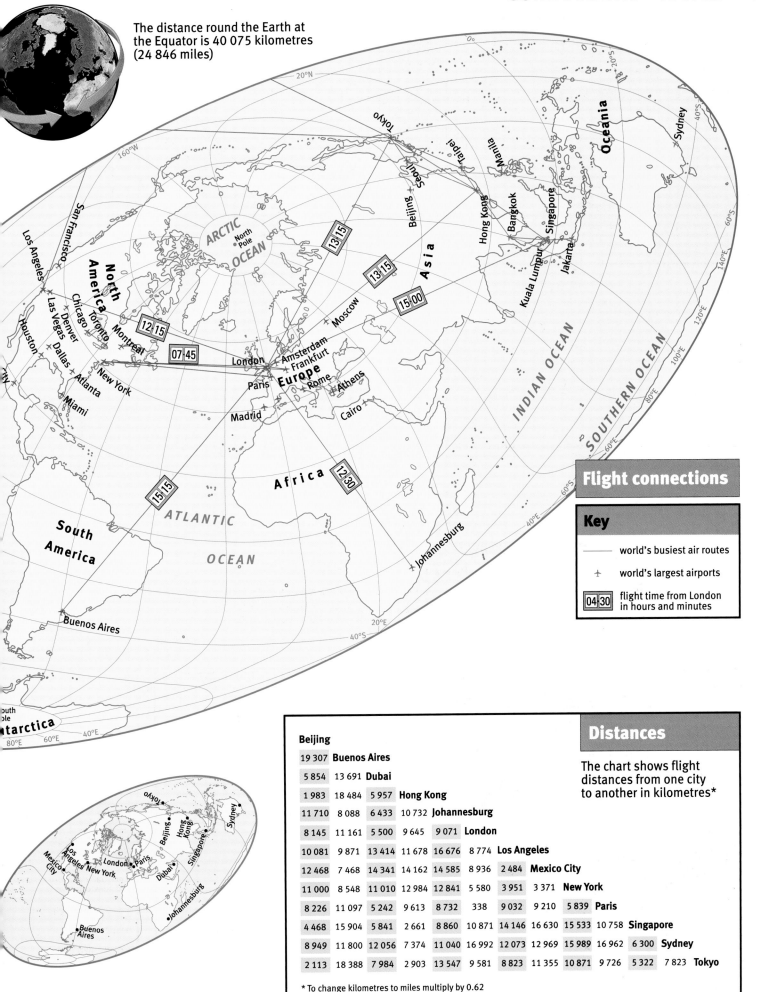

The distance round the Earth at the Equator is 40 075 kilometres (24 846 miles)

Flight connections

Key

——	world's busiest air routes
✈	world's largest airports
04·30	flight time from London in hours and minutes

Distances

The chart shows flight distances from one city to another in kilometres*

Beijing										
19 307	**Buenos Aires**									
5 854	13 691	**Dubai**								
1 983	18 484	5 957	**Hong Kong**							
11 710	8 088	6 433	10 732	**Johannesburg**						
8 145	11 161	5 500	9 645	9 071	**London**					
10 081	9 871	13 414	11 678	16 676	8 774	**Los Angeles**				
12 468	7 468	14 341	14 162	14 585	8 936	2 484	**Mexico City**			
11 000	8 548	11 010	12 984	12 841	5 580	3 951	3 371	**New York**		
8 226	11 097	5 242	9 613	8 732	338	9 032	9 210	5 839	**Paris**	
4 468	15 904	5 841	2 661	8 860	10 871	14 146	16 630	15 533	10 758	**Singapore**
8 949	11 800	12 056	7 374	11 040	16 992	12 073	12 969	15 989	16 962	6 300	**Sydney**
2 113	18 388	7 984	2 903	13 547	9 581	8 823	11 355	10 871	9 726	5 322	7 823	**Tokyo**

* To change kilometres to miles multiply by 0.62

| Country area in square kilometres | Population estimated number of people in 2004 <br> represents 10 million people | Family size number of children in an average family <br> one child

 Years of life number of years people can expect to live <br> represents 10 years | Work if there were 100 people in the country, this is where they would work <br> farms factories <br> offices and services | Rich and poor the average amount each person spends in a year, converted into US dollars
 $1000 $500
 Health the number of doctors for every 10 000 people <br> one doctor |

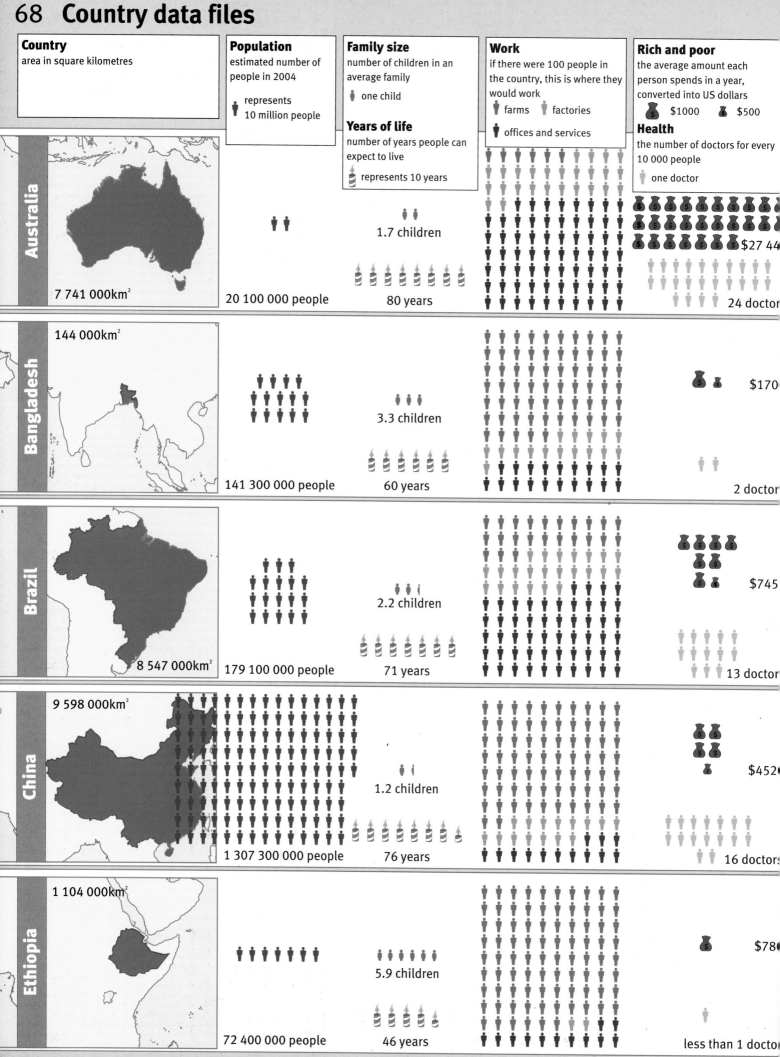

Australia
7 741 000km²
20 100 000 people
1.7 children
80 years
$27 44
24 doctor

Bangladesh
144 000km²
141 300 000 people
3.3 children
60 years
$170
2 doctor

Brazil
8 547 000km²
179 100 000 people
2.2 children
71 years
$745
13 doctor

China
9 598 000km²
1 307 300 000 people
1.2 children
76 years
$452
16 doctors

Ethiopia
1 104 000km²
72 400 000 people
5.9 children
46 years
$78
less than 1 doctor

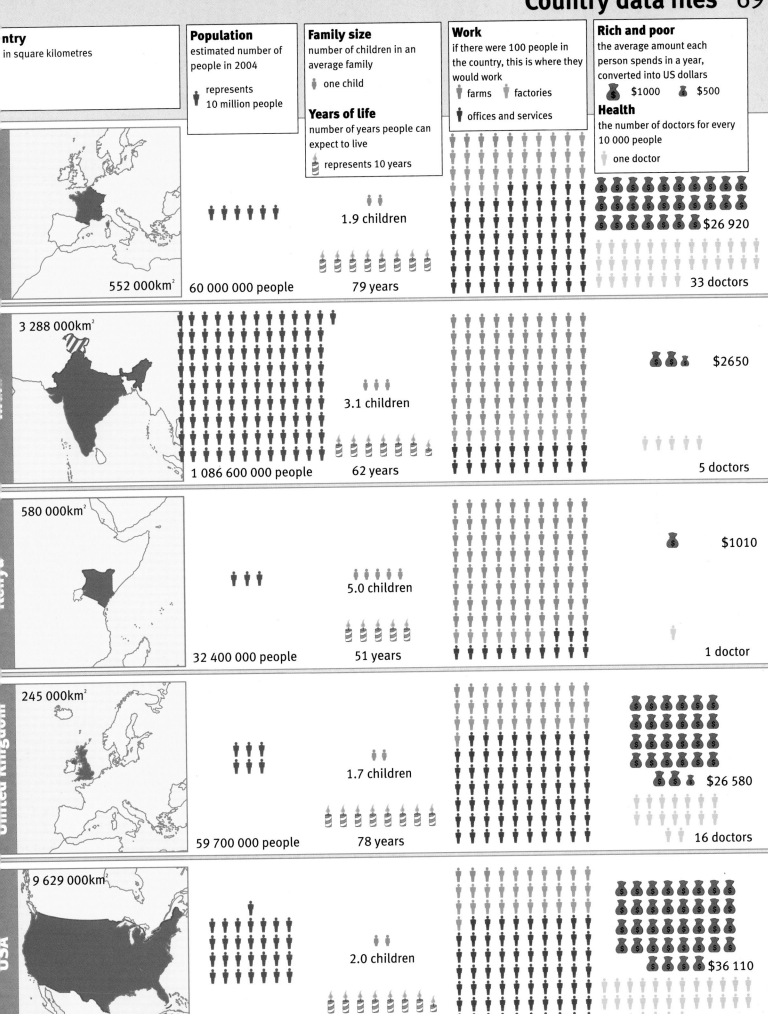

Country
in square kilometres

Population
estimated number of people in 2004

👤 represents 10 million people

Family size
number of children in an average family

👤 one child

Years of life
number of years people can expect to live

🕯 represents 10 years

Work
if there were 100 people in the country, this is where they would work

👤 farms   👤 factories

👤 offices and services

Rich and poor
the average amount each person spends in a year, converted into US dollars

💰 $1000   💰 $500

Health
the number of doctors for every 10 000 people

👤 one doctor

---

552 000km² | 60 000 000 people | 1.9 children · 79 years | | $26 920 · 33 doctors

3 288 000km² | 1 086 600 000 people | 3.1 children · 62 years | | $2650 · 5 doctors

580 000km² | 32 400 000 people | 5.0 children · 51 years | | $1010 · 1 doctor

245 000km² | 59 700 000 people | 1.7 children · 78 years | | $26 580 · 16 doctors

9 629 000km² | 293 600 000 people | 2.0 children · 77 years | | $36 110 · 28 doctors

USA

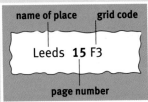

World Flags

 Afghanistan
 Albania
 Algeria
 Andorra
 Angola
 Antigua and Barbuda
 Argentina

 Armenia
 Australia
 Austria
 Azerbaijan
 Bahamas
 Bahrain
 Bangladesh

 Barbados
 Belarus
 Belgium
 Belize
 Benin
 Bhutan
 Bolivia

 Bosnia-Herzegovina
 Botswana
 Brazil
 Brunei
 Bulgaria
 Burkina
 Burundi

 Cambodia
 Cameroon
 Canada
 Cape Verde
 Central African Republic
 Chad
 Chile

 China
 Colombia
 Comoros
 Congo
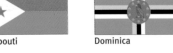 Congo, Dem. Rep.
 Costa Rica
 Côte d'Ivoire

 Croatia
 Cuba
 Cyprus
 Czech Republic
 Denmark
 Djibouti
 Dominica

 Dominican Republic
 East Timor
 Ecuador
 Egypt
 El Salvador
 Equatorial Guinea
 Eritrea

Estonia
Ethiopia
Fiji
Finland
France
French Guiana
Gabon

Gambia
Georgia
Germany
Ghana
Greece
Greenland
Grenada

Guatemala
Guinea
Guinea-Bissau
Guyana
Haiti
Honduras
Hungary

Iceland
India
Indonesia
Iran
Iraq
Ireland
Israel

Italy
Jamaica
Japan
Jordan
Kazakhstan
Kenya
Kiribati

 Kuwait
Kyrgyzstan
Laos
 Latvia
Lebanon
Lesotho
 Liberia

© Oxford University Press

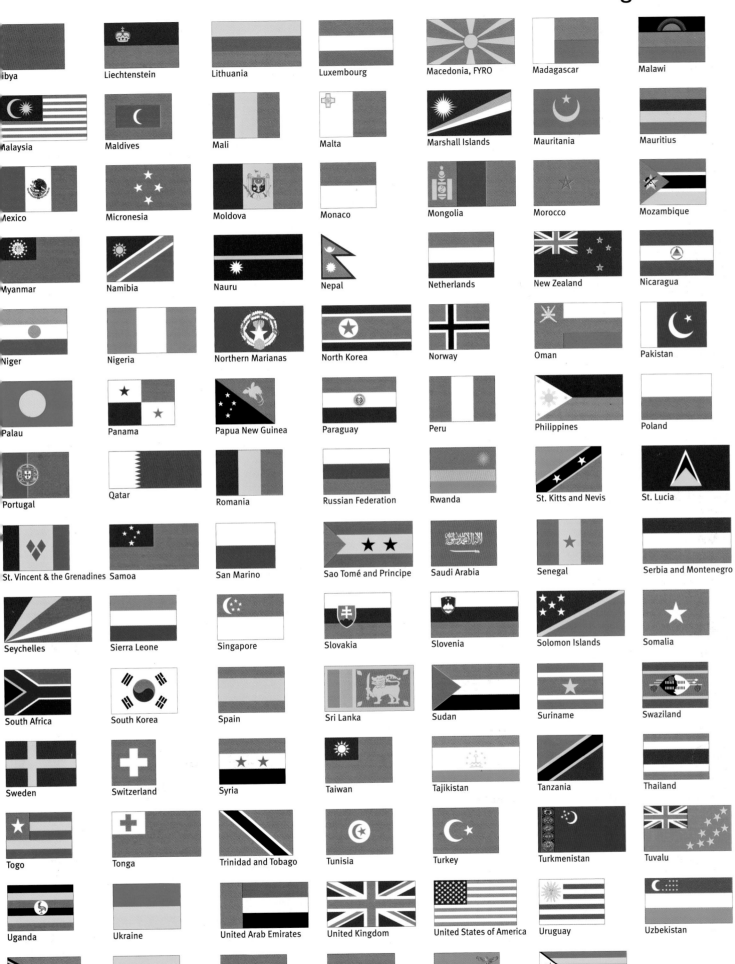

ibya
Liechtenstein
Lithuania
Luxembourg
Macedonia, FYRO
Madagascar
Malawi

Malaysia
Maldives
Mali
Malta
Marshall Islands
Mauritania
Mauritius

Mexico
Micronesia
Moldova
Monaco
Mongolia
Morocco
Mozambique

Myanmar
Namibia
Nauru
Nepal
Netherlands
New Zealand
Nicaragua

Niger
Nigeria
Northern Marianas
North Korea
Norway
Oman
Pakistan

Palau
Panama
Papua New Guinea
Paraguay
Peru
Philippines
Poland

Portugal
Qatar
Romania
Russian Federation
Rwanda
St. Kitts and Nevis
St. Lucia

St. Vincent & the Grenadines   Samoa
San Marino
Sao Tomé and Principe
Saudi Arabia
Senegal
Serbia and Montenegro

Seychelles
Sierra Leone
Singapore
Slovakia
Slovenia
Solomon Islands
Somalia

South Africa
South Korea
Spain
Sri Lanka
Sudan
Suriname
Swaziland

Sweden
Switzerland
Syria
Taiwan
Tajikistan
Tanzania
Thailand

Togo
Tonga
Trinidad and Tobago
Tunisia
Turkey
Turkmenistan
Tuvalu

Uganda
Ukraine
United Arab Emirates
United Kingdom
United States of America
Uruguay
Uzbekistan

Vanuatu
Venezuela
Vietnam
Yemen
Zambia
Zimbabwe